Serious Accidents and Human Factors

Serious Accidents and Human Factors

Breaking the chain of events leading to an accident: lessons learned from the aviation industry

Masako Miyagi

John Wiley & Sons, Ltd

Professional Engineering Publishing

Published by   John Wiley & Sons Ltd, The Atrium, Southern Gate, Chichester,
                    West Sussex PO19 8SQ, England
                    Telephone   (+44) 1243 779777

Email (for orders and customer service enquiries): cs-books@wiley.co.uk
Visit our Home Page on www.wiley.com

Other Wiley Editorial Offices

John Wiley & Sons Inc., 111 River Street, Hoboken, NJ 07030, USA

Jossey-Bass, 989 Market Street, San Francisco, CA 94103-1741, USA

Wiley-VCH Verlag GmbH, Boschstr. 12, D-69469 Weinheim, Germany

John Wiley & Sons Australia Ltd, 33 Park Road, Milton, Queensland 4064, Australia

John Wiley & Sons (Asia) Pte Ltd, 2 Clementi Loop #02-01, Jin Xing Distripark, Singapore
129809

John Wiley & Sons Canada Ltd, 22 Worcester Road, Etobicoke, Ontario, Canada M9W 1L1

Wiley also publishes its books in a variety of electronic formats. Some content that appears in
print may not be available in electronic books.

British Library Cataloguing in Publication Data

A catalogue record for this book is available from the British Library

ISBN 1-86058-473-X

Typeset by Data Standards Ltd, Frome, Somerset, UK
Printed and bound in Great Britain by Antony Rowe Ltd, Chippenham, Wiltshire
This book is printed on acid-free paper responsibly manufactured from sustainable forestry
in which at least two trees are planted for each one used for paper production.

Contents

Acknowledgements

This study, based on data analysis, was made possible through the co-operation and support of many individuals and organizations. With the publication of this book, I would like to express my respect for the professionalism of the incident reporters, and to express my deep appreciation to the late Professor Emeritus Chikio Hayashi, the Toyota Foundation, Senior Commissioning Editor Sheryl Leich of Professional Engineering Publishing Limited, Ann-Marie Halligan and Martin Tribe of John Wiley & Sons, Ltd, and Rodger Williams, who is responsible for Publications Development at the American Institute of Aeronautics and Astronautics. I would also like to offer my sincere thanks to the Secom Science and Technology Foundation and to all of the other individuals and organizations that provided the support, guidance, and co-operation without which this project would not have been possible.

Masako Miyagi
Executive Director, Japan Research Institute of Air Law

List of Figures

List of Tables

Foreword

IRAS: The key to prior prevention of accidents in complex systems

When I look at the work that Ms Miyagi has done, I see in it a reflection of my own dreams. The completion of this book is, in a sense, a realization for me of those dreams.

Allow me to go back some 30 years or so. At that time, I was involved in accident prevention research with the former Japan National Railway. My original intention had been to calculate the 'probability of error', and, using this as a base, to work out a method for calculating the degree of danger in various specific types of accident occurrence, from the perspective of probability theory. Examining the data, however, I came to feel that this approach would not be useful in developing actual accident prevention measures.

An accident is one type of phenomenon – one that occurs as a result of a chain of errors. I was aware that, if this chain could be broken at some point, then the accident would not occur. I thus reached the conclusion that if one were to conduct a comparative analysis of 'the chain of errors seen as a phenomenon' and 'accidents that did not occur because the chain was broken at some midway point' – a situation referred to in this text as an 'incident' – then it might be possible to discover the roots of an approach to accident prevention. I planned to make this investigation the next stage of my research. The labour union of the former National Railway was quite strong, however, and the authorities met with opposition to the implementation of such an investigation. It was a truly regrettable situation.

Some years later, when I was working as a member of the Toyota Foundation's committee for the selection of recipients for research grants, I learned of Ms Miyagi's work in the field of Incident Reports Analysing Systems (IRAS), and was astonished by the similarity with my own research. Although the study saw some measure of resistance, it gained the support of most of the committee members, and research began in earnest in 1984. On seeing the initial results, the entire committee was most impressed, and as a result the grants continued to be offered, with the unanimous consent of all members, on a number of occasions up to 1995. Such a situation was unprecedented in the history of the Foundation.

The extremely detailed final results have been published by Yuhikaku Publishing Co., Ltd, in three extensive volumes. The current text contains the essence of these volumes, summarized and presented in an easy-to-read format.

There is something enthralling about this theoretical method, a method that investigates incident reports from the perspective of clear and penetrating analysis, and uses quantitative analysis to draw to the surface the critical accident information that is hidden among the results. I have the greatest admiration for the sharp vision that can see through to the true nature of the accident – that it is a chain of component elements – and attempt to bring these elements into focus through the use of quantification method III. This is a wonderful effort, something that I have rarely seen in all the statistical research that I have come across. And it is rarer still to find such a determined research stance, which does not stop at theoretical research, but rather continues on to promote the development of the network of trust relationships that is required to gather information on the various incidents.

This type of approach is extremely effective, not only in the field of aircraft accidents but also in relation to all complex systems and to atomic power plants in particular. I shall be most pleased if this text leads to the creation of an IRAS organization, which will become the 'trusted neutral organization' that Ms Miyagi has long wished for.

Chikio Hayashi
Professor Emeritus and Former Chief
Institute of Statistical Mathematics

Introduction

This introduction will present some background regarding the Incident Reports Analysing System (IRAS), which is the central issue of this book, so that the reader may gain a deeper understanding of the basic philosophy and the specialized aspects of this system. I will focus specifically on Multidimensional Analysis of Incident Reports (MAIR), which was developed by the author as a method of analysing individual incidents involving human factors,[1] on the need for analysis from a comprehensive perspective in this area, and on the reasons for using quantification method III to achieve this goal.

There is little doubt that, regardless of the field, safety is one of the most critical issues that will be faced in the twenty-first century.

In recent years, progress and development in science and technology have made dramatic contributions to human society. However, these same developments have given rise to many new types of danger, and a massive increase in losses that would have been inconceivable in the past. Another effect of these changes can be seen in a shift in the causes of accidents from 'hard' factors to 'soft' factors (see Fig. 1)[2] – in other words, in proportion to the speed of technological reforms, human causes have come to be emphasized more than mechanical causes (see Fig. 2).

This trend is by no means an indication of carelessness on the part of the individuals involved; rather, it could be considered an indication that the methods used to implement traditional safety measures in the past have reached a limit of effectiveness. This is because the most basic safety measures taken in the past were limited to reprimands and punishments targeting the persons responsible for the accidents, and improvements to mechanical aspects stemming from the results of accident investigations. Such accident investigations placed an emphasis on technical analysis of events in accidents that had already occurred, and for this reason there is no question that they contributed

ICAO Human Factors Training Manual

About three out of every four accidents ($\neq$75%)

According to U.S. FAA survey (1980)

92% of total-loss accidents

According to the Marine & Fire Insurance Association of Japan*

Missile accidents: 20–53%
Electronic equipment related accidents: 50–70%

* Prevention Report No. 143, p.23

Fig. 1 Ratio of accidents involving human factors

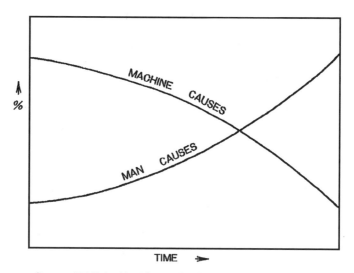

Source: ICAO Accident Prevention Manual (1984 p.11 Fig. 3)

Fig. 2 Trends in movement of accident causes, the shift from machine causes to man causes

to a sharing of important information regarding the mechanical aspects of these accidents, that this information was put to use in making improvements, and that significant results were achieved through this process.

Human beings are able to develop and increase their abilities to some extent through education and training. The fact remains, however, that it is extremely difficult to obtain the information on human aspects of accidents that would be required to implement such training, because the people most directly involved may have been killed in the accident, or may be reluctant to come forward for fear of being held responsible. There are definitive limitations to the approach described above even if all the relevant information is obtained; namely that when studies are made into accident prevention measures based on accident investigations, the investigations can only begin after the accident has occurred. Furthermore, the improvement measures based on accident investigations will only be of value in preventing the re-occurrence of accidents that are identical to those on which the measures were originally based.

Looking back, if we consider the structure of an unsafe event (an accident or incident), we can see that, with the exception of intentional acts or those involving mental disorders, various factors of differing dimensions set the stage for human error, intertwining in complex patterns and forming a chain along a time line that leads to the occurrence of the accident or incident. In many cases, however, intervening factors come into play at some point along that chain, such that the accident is averted.[3] As indicated by Heinrich's law,[4] there is no question that, before a single major accident occurs, many small accidents or similar unsafe events (referred to in this book as 'incidents') have occurred before on a large or small scale, but have not resulted in major accidents.

One example of this pattern can be seen in the case of the midair collision near the German border on 1 July 2002,[5] on a route that crossed Swiss Air Traffic Control space, as an extremely similar serious incident (a near miss) had occurred in Japan in the previous year, on 31 January 2001.[6] In both the midair collision over Germany and the near-miss case in Japan, all of the aircraft involved were equipped with a Traffic Alert and Collision Avoidance System (TCAS), and in each case the system was operating normally. Both cases occurred when one of the approaching aircraft descended in spite of resolution advisories (RAs) from the TCAS to avoid collision by ascending. These cases generated much debate regarding the fact that it is not necessarily clear whether the pilot should give precedence to ATC instructions or TCAS RAs when making decisions in the event that these instructions are in conflict.

In order to prevent accidents before they occur, and to respond to the new types of danger that have arisen as a result of technological innovations, it is therefore necessary to change our approach from the 'retrospective' standpoint based on accident investigations that has been adopted in the past to the establishment of more 'prospective' measures.

I believe that the establishment of IRAS, as described in this book, will play a critical role in making this transition. This is because IRAS serves the function of seeking out similar factors that come together to form major accidents from among large numbers of incident examples, and facilitating an understanding, from a comprehensive perspective, of the correlation between the background factors that led up to the relevant human errors and the form that those errors take. Furthermore, it contributes to the process of determining which factors among the potential risk factors in a given field represent the greatest dangers and thus require the immediate implementation of improvement measures.

In other words, the goal of IRAS is not to attempt to secure safety through reprimands and punishments targeting individuals guilty of error.[7] Rather, it begins by seeking out the conditions (the background factors) that will shed light on why the individual made those errors and gaining an objective understanding of those conditions from a comprehensive perspective, so as to contribute to the prevention of accidents before they occur.

Up until 1975, the rate of occurrence of fatal accidents involving international civil airline flights (with the exception of the USSR) had been on the decline; this occurrence rate demonstrated a tendency to either level off or increase after 1975 (see Fig. 3). Expressing concern for these statistics, and based on an awareness of the limitations presented by traditional approaches to safety measures, the International Civil Aviation Organization (ICAO) published an Accident Prevention Manual (APM) in 1984 with the intention of urging various countries to study incident reporting systems as a new approach to this issue.

In 1980, the Japan Research Institute of Air Law, for which the author is the Executive Director, had already presented to the Japanese Minister of Transport and other authorities a 'Proposal for the prevention of aircraft accidents',[8] pointing out the need for an incident reporting system. Later, with the assistance of the Airline Pilots Association – Japan (ALPA-J), the Institute conducted an extensive survey[9] of reporting systems throughout the world, including NASA's Aviation Safety Reporting System (ASRS, United States), Australia's air safety incident reporting, the United Kingdom's mandatory occurrence reporting system, Sweden's hazard reporting system, Finland's air miss report, and six reports originating in New Zealand. Based on the results of this survey, it then presented a 'Proposal for the securing of aviation safety – a request for the establishment of a Total Air Safety System',[10] the contents of which included reporting methods

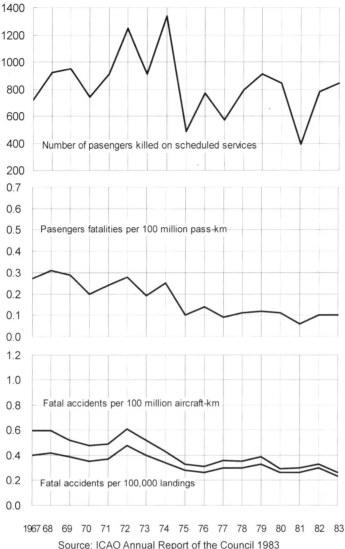

Fig. 3 Fatal accident rate for passengers on world scheduled services 1964–1983

and organizational requirements for the implementation of IRS, among other elements. The fundamental principle of this proposal was that the key requirements for such an organization are practical independence and comprehensiveness, and that the key requirements for reporting methods are anonymity and a free essay format.

Since 1984, with research assistance from The Toyota Foundation, the Japan Research Institute of Air Law has initiated independent actual studies targeting aircraft operations, maintenance, and air traffic control respectively. In each case, it independently gathered incident reports in free essay format relating to human factors, eventually obtaining a total of 1326 highly accurate responses. (The incident examples referred to in this book were gathered using this type of method.) The Institute also developed a Multidimensional Analysis of Incident Reports (MAIR) as a method of analysing individual incidents, and conducted analysis from a comprehensive perspective using the statistical method discussed below.

In the past, a number of methods have been put forward for the analysis of unsafe events related to human factors, including 4M,[11] 6P,[12] SHEL,[13] and fault tree analysis.[14] While 4M, 6P, and SHEL are meaningful in terms of categorizing analysis results, they cannot be considered appropriate for clarifying the structure of unsafe events as a chain of events on a time line, in which a variety of background factors interact with elements of human error. Meanwhile, although fault tree analysis may be effective in the analysis of individual mechanical failures, there are numerous problems with applying this method to this particular field. In order to analyse human factors and their background factors, it would be necessary to gather large volumes of information on individual cases, and follow-up studies would be difficult in the case of systems based on the principle of anonymity. In addition, because of the very short space of time between the point at which the individual realizes that an incident has occurred and the action taken to recover normal operations, even if the person taking that action has a clear recollection of the important elements that led up to the incident in question, persons who were not directly involved when the incident occurred cannot be expected to have such a clear recollection, and the recollections they do have may be incompatible with the actual circumstances.[15] Finally, there are many cases in which it is inappropriate to limit the characteristics of human error or inefficiency into two extremes of 'right' and 'wrong'.

Based on an awareness of the above problems, we developed a new analysis premised on the establishment of IRAS in order to analyse individual incidents related to human factors. MAIR is the result of these efforts. (See page 46 onwards for a more detailed description of MAIR.)

MAIR looks at every type of task as a series of information processing activities made up of four repeating phases: 'information source', 'information receiving', 'decision', and 'action or instruction'. Each single 'event' is defined by a different information source or a different individual performing the action. An unsafe event (accident or incident) is made up of one or more of these events in which a flaw exists in any of the four phases.

We have designated nine categories to define the various types of relationship that might exist between the target at which physical or psychological energy should have been directed ('what was required') and the target at which it was actually directed ('what was actually done') in the context of human information processing activities. Applying each of these categories to the three relevant phases of information processing activities – 'information receiving', 'decision', and 'action or instruction' – results in a total of 27 'modes' (see Table 1 on page 50).

When an unsafe event occurs, the path in which physical and psychological energy is directed may be affected by a variety of background factors. Even in the same event, the three phases will not necessarily correspond to the same categories. For this reason, it is necessary to analyse what background factors were present for each event, and which categories corresponded to the movement of human energy. For example, even if the information is received correctly, there may be an error in the decision-making or operation/instruction; conversely, even if the correct decision was made in spite of lack of clarity in the information receiving phase, the operation may end up being delayed. Based on the premise of such an analysis, through analysis from a comprehensive perspective, we will be able to determine the probability that a given background factor will give rise to a given distortion or deficiency in human information processing activities.

Other factors to be analysed include: the events that contribute most to the occurrence of the unsafe event in question and the phase in which that event occurs (the incident critical event and phase); the degree to which the incident approaches the stage of an accident under specific conditions of a given unsafe event (i.e. the degree of danger, rather than the degree of breach of duty); the reasons that the incident was prevented from developing into an accident (reasons for incident recovery, or reasons for the break in the chain of events).

It is essential that the subjectivity and preconceptions of the analyst are excluded, and that the background factors used in the analysis are based only on the facts recorded in the report.[16] The analyst therefore must not assume in advance that any factors exist, as there may be cases in which differences arise in the background factors depending on the time at which the survey was conducted. For the purposes of easier management, background factors are divided into the following main

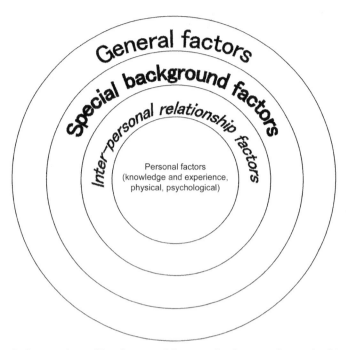

Fig. 4 Categories of background factors for human-factor incidents

categories: personal factors (factors related to knowledge and experience, physical factors, and psychological factors), interpersonal relationship factors, general factors, and special background factors/special circumstances (see Fig. 4). The concrete factors derived from each individual case are input one by one into the main categories described above, and an analysis code table[17] is created. (Which of the four categories to which a given factor applies is not an issue, as each of these factors will be handled separately in parallel in a later analysis using quantification method III.) When the same factor is extracted from different events or different incidents, the same number as assigned for the previous analysis is entered in the column for the appropriate item in the analysis mat.

In conducting an analysis from a comprehensive perspective of the results of analyses of individual incidents using the MAIR method discussed above, we have adopted Hayashi's quantification method III, a statistical method of multidimensional analysis.

Examination of frequency is effective when only handling certain specified factors. Incidents do not occur as a result of single factors, however, but comprise human errors and the factors that lead up to those errors. As such, we will be unable to grasp the true state of affairs regarding potential risk factors in a given analysis without clarifying the interrelationships of these factors from a comprehensive perspective. Factors with a high degree of occurrence, or factors that make up major accidents, are not necessarily those that represent the greatest degree of danger among the potential risk factors in any given field.

Quantification method III was developed in 1956 by Chikio Hayashi, Professor Emeritus and former Chief of the Institute of Statistical Mathematics.[18] In 1973, an absolutely identical method was conceived independently as a corresponding theory by J. P. Benzecri of University of Paris VI.[19]

This is the first time that this method has been implemented in the field of safety, mainly because in the past it has not been possible to gather a sufficient number of incident reports, precluding the development of an appropriate analysis method for individual incidents. It has, however, been applied to various societal studies in Japan, France, and other countries, and its position has already been secured in the field of statistical sciences. References to quantification method III, as well as equations and reference materials, have been quoted from texts by Chikio Hayashi. Please refer to Appendix A for further details.

The main reasons that quantification method III was adopted for this process of analysis from a comprehensive perspective are as follows:

1. Because all the factors contributing to the occurrence of an unsafe event are qualitative, it is necessary to quantify these factors using the concept of the correlation coefficient in order to clarify the relevant interrelationships. Quantification method III is a method for quantifying qualitative factors with no common external standards, and clarifying the correlations among these factors.

   For example, in terms of the operation of an aircraft, let us look at the following factors: Early morning' (one of the items in 'Time of occurrence'), 'Landing' (in the 'Flight phase'), 'Fatigue' (one of the items under 'Physiological factors'), and 'Delay' (one mode of 'Error'). Each of these is a qualitative factor, and there is no way to define a common, uniform, external standard, so it is impossible to express these factors quantitatively. Nevertheless, there are cases in which correlations can be seen among these individual factors. For example, if we include an additional element to these factors – the specialized condition of 'long-distance flight/time difference' – then, as a result of fatigue from a long-distance flight when the pilot has been flying through the night, when the aircraft is landing in the early morning, the crew in the cockpit will tend to be in a state of 'interdependence'. This will in turn cause delays in information receiving, decision-making, and operation/instruction, possibly resulting in the occurrence of an incident. In this way, we can see that there are cases in which incidents occur as a result of random conflict among various qualitative factors of differing dimensions.

As such, quantification method III is extremely valuable in the context of explorative data analysis. Expressed in the specialized terms of symbolic logic, this method derives the 'eigensolution' for latent equations that use as their coefficients the degree to which factors exist simultaneously. Stated in very simple terms, when these factors are distributed on a graph, similar factors (those occurring simultaneously or with common or similar factors) are placed closer together, while dissimilar factors are placed further apart.

2. As mentioned, quantification method III is a statistical method for multidimensional analysis, and as such the analysis results are positioned in a multidimensional Euclidean space.[20] The factors on the first dimension show the strongest correlation, followed by factors on the second and third dimensions. In Euclidean terms, this means that the largest eigenvalues (the generalized correlation ratios) are computed on the first dimension, the second largest eigenvalues are computed on the second dimension, and the third largest on the third dimension. For this reason, the first dimension has the strongest discrimination, and the factors with the weakest correlativity are positioned at the two extremes of this dimension. By looking closely at the characteristics of the groups of factors at these extremes, we are able to learn the two major characteristics of the potential risk factors in the field in question.[21]

3. If we are able to find the axis on which the factors that express the degree of danger ('serious', 'considerable', 'fairly little', and 'virtually none') are distributed one by one according to the degree of danger, then we can accurately grasp that the degree of danger for various factors is expressed according to the value computed for each of the factors (that is, the categorical value[22] referred to in quantification method III). In this way, through an analysis from a comprehensive perspective of the many potential factors in a given field, we can see which factors present the greatest degree of danger.

4. Factors that occur simultaneously have a strong mutual correlation and are distributed close together, so it is possible to find typical incident types based on empirical observation.[23]

5. The accuracy of the analysis is expressed by values for each dimension (the square root of the eigenvalue),[24] so analysis results

are clearly and objectively expressed in a three-dimensional Euclidean space, enabling a consistent awareness of the facts regarding the many potential risk factors in a given field, even though perspectives may differ. This is also extremely significant in that it facilitates studies of relevant improvement measures.[25]

The first step in applying quantification method III to conduct an analysis from a comprehensive perspective of analysis results for each individual incident (derived by using MAIR) is to calculate the number of relevant items for each category. If the number of relevant items is less than 2–3 per cent of the total number of sampled events, then similar factors are grouped together; in cases where there are no such similar factors, the events are considered independently as special cases, and quantification method III is not applied. For example, the psychological factors of 'impatience' and 'fluster' are similar, so they are grouped together, creating the category 'impatience/fluster'. There was only one relevant case of the physical factor of 'low oxygen', so this case was considered independently.

While a larger number of samples (pieces of data) will generally enable better solutions, as a rule, stable solutions can be obtained if 200 or more samples are available, with the exception of special factors that occur very infrequently. In contrast to the number of samples, however, a greater total number of factors will result in a more unstable solution. It is thus preferable to limit the total number of categories to well below the number of samples.

Once the above process has been completed, all factors to be incorporated into quantification method III are numbered consecutively (referred to as categorical numbers in quantification method III); after organizing each factor for each individual event on the basis of whether or not there is a reaction, the data are input for computation.

Categorical values for each factor (referred to as 'category' in quantification method III) based on quantification method III can be determined easily by a computer using Program Package for Social Science II. The work of interpreting the results of this computation should be left to a qualified analyst.

Unsatisfied with the approach of incident reporting systems that have been commonly advocated in the past as a means of preventing accidents before they occur, the author has added the term 'analysing' to create a new approach – that of the Incident Reports Analysing System. The reason for this is the firm belief that, only through the implementation of MAIR, an original method developed by the author for analysing individual incidents, and quantification method III, which enables analysis from a comprehensive perspective, will it be possible to clarify the essential structure of incidents, and to grasp fully the truth of potential risk factors in a given field.

If a system goes no further than examining the details of individual incidents and feedback related to those incidents, then they are no more than extensions of conventional safety measures based on accident investigations, even if the situation before the accident differs from that following the accident. Safety measures designed to prevent accidents before they occur must be implemented from a wide-ranging perspective based on the actual circumstances regarding risk factors.

Based on the fundamental principles of IRAS as outlined above, the author has gathered incident reports related to human factors in the field of aviation, applying the analysis method MAIR – which was developed to clarify on a time line the mechanisms related to the occurrence of individual incidents involving human factors – and using quantification method III to conduct a comprehensive study of the results of this analysis. This book is a summary of the findings of this actual study based on data analysis.

Regardless of the field or the technologies involved, human error is related to the occurrence of unsafe events. In other words, human error is a common factor. Before a serious accident occurs, there are always incidents that have the same factors. The IRAS method examines information from many incidents and considers the correlation between human error and the background factors that give rise to human error as an overall picture from the perspective of the degree of danger. This will contribute to preventing accidents in complex large-scale systems before they occur, not only in the single field of aviation, but in many other fields as well, including astronautics, atomic power, genetics, and medicine.

Notes

1 The concept of 'human factors' has not yet been clearly defined (it is still in flux), but human factors should be clearly delineated from the phenomenon of 'human error', and from the concept of 'negligence', in which legal valuations are included in human error. In this book, the term 'human factors' is used when examining human-related factors in cases where incidents or accidents occur in relation to the conditions (background factors) that bring about human errors.

   Incidentally, at the start of a technical conference of the International Air Transport Association (IATA) held in Istanbul in November 1975, the following confirmation was presented. In the not so distant past, there was a tendency to classify human-factor accidents as 'pilot error' and leave the matter at that. This classification carried with it the connotation of culpable fault on the part of the flight crews and had two very serious consequences. Firstly, there was a tendency on the part of accident investigators and airline management to write off 'pilot error' accidents as being solved; secondly, it led to a defensive attitude of confrontation on the part of the pilot group in attempting to deal with these problems. The conference marked a clear recognition that, with very rare exceptions, 'pilot error' accidents are not due to negligence or deliberate misbehaviour by the crews, but are rather the byproduct of a series of circumstances that put the crews in a position where the probability of them making an error was extraordinarily high. All three parties – airlines, governments, and pilots – are now starting to ask the critical question of why such errors are made. As the industry goes more deeply into this subject, the concept of fault is disappearing, pilot/management relations are becoming more objective and co-operative, and the prospect of preventing a large percentage of these accidents is opening up.

Chapter 1 of ICAO Digest No. 1, CAP719 *Fundamental Human Factors Concepts* also discusses the meaning of human factors.

2 In the forward to *Human Factors Training Manual* (Doc 9683-AN/950) published by the International Civil Aviation Organization (ICAO), it is noted that 'It has long been known that some three out of four accidents result from less than optimum human performance'. Furthermore, according to a US FAA survey, human factors related to all persons involved in aircraft operations (crew, maintenance personnel, controllers, etc.) are said to be involved in 92 per cent of all fatal accidents and 62 per cent of all monetary losses. (Conner, T. M. and Hamilton, C. W., Evaluation of safety programs with report to the cause of air carrier accident. DOT, FAA, Washington, DC, Report No. ASP80-1, January, 1980.)

Human factors are also said to account for between 20 per cent and 53 per cent of missile accidents, and between 50 per cent and 70 per cent of accidents related to nuclear facilities (Yobou Jihou [Prevention Report] Vol. 143, p. 23, Tokyo: General Insurance Association of Japan).

3 Refer to Fig. 6 on page 17.

4 Refer to Fig. 5 on page 16.

5 Bundesstell für Flugunfalluntersuching (2002) AX001-V-2/02 State Reports, Domogala, P., Ruitenberg, B., and Stock, C. (2002), The mid-air collision itself: the facts. *The Controller*, Vol. 41, p. 61 onwards. IFATCA, in English.

Iwamoto, K. (2003), Doitsu jouku de TCAS tosai-ki doushi ga kuuchuu shototsu [Two aircraft equipped with TCAS collided in midair near

Germany]. *All-Nippon Airways (ANA) Flight Safety Review*, 2003, No. 221, p. 24 onwards, in Japanese.

The following is a summary of that accident.

On 1 July 2002, Bashkirian Airlines charter flight BTC2937 (a Tupolev 154), which had departed from Moscow and was bound for Barcelona via Munich, collided with DHL flight DHX611 (a Boeing 757 cargo plane), which was flying from Bahrain via Bergamo, Italy, to Brussels. The collision occurred at a flight level of approximately 35 000 ft, above Lake Constance, which is located near Ueberlingen, Germany, close to the Swiss border. A total of 71 people were killed, including passengers and crew. Both aircraft were equipped with the same Traffic alert and Collision Avoidance System (TCAS; Honeywell 2000). Each TCAS issued collision warning instructions (Resolution Advisories, or RA); the Boeing was instructed to descend, and the Tupolev was instructed to climb, although this conflicted with the Swiss Air Traffic Controller's instructions to descend. Regardless of the TCAS RA, the Tupolev followed the ATC instructions, and proceeded to decrease its altitude. The crew of the Boeing received a further RA from the TCAS system to increase their rate of descent, so as to avoid the now descending Tupolev, but just over 20 s later the two aircraft collided at a roughly 90° angle.

Results of an investigation provided the following information. The Short-Term Conflict Alert (STCA), which indicates dangerous approaches of two aircraft in the same altitudes, was not operating at the Zurich area control centre. The air control centre for the adjacent airspace in Karlsruhe was unable to make direct contact by phone. The controllers at the Karlsruhe upper area control centre had been advised of the danger of collision 2 min ahead of time through their own STCA, and attempted to contact the Zurich ACC, but were unable because the lines were busy. Furthermore, despite operational guidelines to the contrary, one of the two air traffic controllers on duty at the Zurich ACC was on a break, leaving only one controller monitoring two radar screens.

6 Ministry of Land, Infrastructure and Transport Japan, Aircraft and Railway Accidents Investigation Commission, 2002 Aircraft Accident Report, 2002-5; Watanabe, Y. (2002), Yaizu-oki de JAL-ki ga near-miss [JAL flight was involved in a near miss over the water off Yaizu]. *ANA Flight Safety Review*, 2002, No. 219, p. 18 onwards. The following is a summary of that case.

On 31 January 2001, Japan Airlines (JAL) flight 907, a B-747-400 departing from the Tokyo International Airport in Haneda and bound for Naha Airport in Okinawa, and JAL flight 958, a DC-10 that had departed from Pusan International Airport bound for New Tokyo International Airport in Narita, were involved in a near-miss incident at a flight level of approximately 35 000 ft, on flight paths that crossed above Yaizu, 106 miles south-west of Tokyo. Nine of the passengers and crew on the 747 suffered serious injuries, with 91 suffering minor injuries. A part of the aircraft was also slightly damaged. No injuries were suffered by the 250 persons on board the DC-10, and the aircraft itself was not damaged. Both aircraft were equipped with TCAS, which operated correctly, instructing the 747 to climb and the DC-10 to descend. Before receiving the TCAS advisory, however, the 747 had received instructions from the air traffic controller to descend as well, resulting in the near miss.

Like the previously mentioned collision over Germany, this near miss raised the issue of whether ATC instructions or TCAS instructions should take precedence in the event that these instructions are in conflict.

Because the instruction advising the JAL 907 flight to descend was given by a controller in training, there was a controller overseeing that trainee's actions. In May 2003, the Tokyo Metropolitan Police Department had these two air traffic controllers and the captain of JAL flight 907 turned over to prosecution on charges of professional negligence.

The Nihon Keizai Shimbun, 31 March 2004.

'On March 30, 2004, The Tokyo Public Prosecutors Office indicted two air traffic controllers on charges of professional negligence resulting in bodily injury, but the Captain of flight JAL907 was not prosecuted. The charge of negligence on the part of the Captain was rejected on the gounds that, according to regulations at the time, in the event of a conflict between TCAS resolution advisories (RAs) and ATC instructions, the decision as to which should be followed is left to the discretion of the Captain. The Ministry of Land, Infrastructure and Transportation Japan has issued thorough directions to all airline companies, that in the event of a conflict between TCAS RAs and ATC instructions, the TCAS RAs should be followed.'

7 This is not to say, of course, that reprimands and punishments for the persons involved are not required; the principle of IRAS is that the problem of determining responsibility should be considered as separate from the perspective of ensuring safety.

8 Japan Research Institute of Air Law (1980), Kokujiko-boshi no tame no Teigen [Proposals for the prevention of airline accidents] (in Japanese). *Review of Air Law and its Practice*, 1980, No. 13, p. 5 onwards. Tokyo: Yuhikaku Publishing Co., Ltd.

9 Sasada, N. (1981) Kakukoku no Incident Reporting System nituite [Incident reporting systems in various states] (in Japanese). *Review of Air Law and its Practice*, 1981, No. 13, p. 37 onwards and No. 14, p. 5 onwards. Tokyo: Yuhikaku Publishing Co., Ltd.

Sakai, I. (1983), Kuni niyoru Incident Reporting System [Incident reporting systems on the national level] (in Japanese). *Review of Air Law and its Practice*, No. 15, p. 35 onwards.

10 Japan Research Institute of Air Law (1983), Koku no Anzen Kakuho notame no Teigen – Total Air Safety Kakuritsu no yosei [Proposals for

ensuring aviation safety – a request for the establishment of total air safety] (in Japanese). *Review of Air Law and its Practice*, 1983, No. 15, p. 35 onwards.

11 Proposed in 1968 by C.O. Miller, who suggested that accident factors develop from factors symbolized by one of four categories – man, machine, media, and management – or through competition among these factors. In the APM published by ICAO in 1984, media was replaced with environment, and management with media, but the basic details of the approach are much the same: that 'Unsafe elements can be divided into the categories of Man, Machine, and Environment, and an unsafe flight refers to the relationship between these three basic elements and the Mission' (ICAO APM 3.1.20 and 3.1.22).

12 Advocated by A.F. Zeller. This approach suggests that the key factors affecting the human element of safety are physiological factors, psychological factors, psychosocial factors, physical factors, and pathological factors. A related proposal adds a sixth 'P', pharmaceutical factors. The 5P method was used by the American military to analyse two incidents of speed loss in C-5 transport aircraft in 1984 (Majors, J.B., *Human Factors Survey: C-5 Pilots*, USAFSAM-TR-84–26, 1984).

13 Advocated by E. Edwards (1972) and developed by F. Hawkins (1975). SHEL refers to Software, Hardware, Environment, and Liveware.

14 Ref. MIL-STD-882, 'System Safety Programme for System and Associated Subsystem and Equipment'. ICAO APM 4.3.8 shows an example of a simple fault tree.

15 Because the reporters are professionals in their respective fields, however, important items related to the occurrence of the incidents are clearly remembered, even if the incident occurred over a very short period of time. This is demonstrated by the high analysis accuracy of quantification method III, which will be discussed later. Unlike investigations in non-specialized areas of society, in the case of specialized fields, if the reporters report their experiences honestly, one can expect that a sufficient amount of information will be obtained.

16 The difference between the actual facts of an incident and analysis based on the preconceptions of a third party are clearly demonstrated in the differing results for analysis of 'occurrence' and 'discovery' related to aircraft maintenance (Ref. 'The great difference between occurrence and discovery'; p. 353 onwards).

17 The analysis code table is created so as to allow entry of the factor items shown in Table 2 (aircraft operation) on page 111, Table 3 (ATC) on page 146, and Tables 4 and 5 (aircraft maintenance) on pages 200 and 203.

18 Hayashi, C. (1956), Theory and example of quantification (II) (in Japanese). *Proceedings of the Institute of Statistical Mathematics* 4(2): pp. 19–30.

19 Benzecri, J. P., *et al.* (1973), *L'Analyse des Données 1, 2* (in French). Paris: Dunod.

20 Ref. Fig. 15 on page 122.

21 The actual concepts for axes used in the descriptive representation of data are not applied to the axes displayed in a multidimensional analysis such as quantification method III. Each axis, generally, represents the sort of observation that has been measured for a subject (e.g. the X axis represents the height, and the Y axis represents the weight for a person). By plotting specific figures of subjects on the X–Y (two-dimensional) space, we understand the relationship between the values of two sorts of observations.

The axes displayed in the quantification method III analysis do not express any sort of observations. In the analysis, the new values for the observations and the subjects or factors are computed synthetically so that the subjects or factors showing strong correlations between the measurements are placed in close proximity. The X axis represents the most suitable variable for this purpose. The variable of the Y axis is computed so that closeness is represented on the two-dimensional space, and so on.

The axes thus play the role of a scale showing the interrelationships between the various subjects or factors. At the same time, because as a result it is possible to see the standards by which all of the subjects or factors have been classified, by taking into consideration the conditions of the distribution of the various subjects or factors, we can derive the meanings expressed by the axes that separate all of these subjects or factors.

22 The categorical value referred to in quantification method III; that is, the value computed for each of the factors in this book.

23 This is referred to as a 'cluster'. Incidents occurring in reality do not always correspond to the factors that make up each cluster. In some cases, one factor may be missing; in others, a different factor may be involved. Even so, the factors that make up each cluster are the most typical factors comprising the type of incident in question, and it thus becomes easier to judge whether it will be effective to study or implement improvement measures targeting those factors (refer to Fig. 13 on page 118).

24 The square roots of the eigenvalues differ depending on the accuracy of the data, but it has been learned through experience that meaningful information can be derived when these values are roughly between 0.6 and 0.3. In standard societal surveys, the square roots of the eigenvalues for the first dimension are around 0.5, and for the second and third dimensions they fall to around 0.3.

25 Even in the aviation field, a wide range of improvement measures has been proposed from a variety of perspectives. One cannot deny, however, that none of these measures has been effectively implemented because of an inconsistent awareness of the facts resulting from these differing perspectives.

Chapter 1

Presence of accident warning signs in all incidents

1.1 The great losses that result from accidents in large-scale systems

Spectacular progress in technology has gradually been making reality of the boundless dreams of mankind, and has greatly improved our lives. However, that same progress has brought to human society entirely new types of disaster that were previously unimaginable to us. Particularly in cutting-edge fields of technology such as aeronautics, astronautics, atomic power, and genetics, systems are increasing in complexity and scale. This in turn has brought about a dramatic increase in the danger of an accident resulting in immeasurable loss.

In fields such as these, the human and economical loss that results from an accident is not limited to the company or organization directly involved in the accident. Rather, these losses may extend to unrelated third parties and even to the level of the national economy.

In the case of the Chernobyl nuclear power plant accident, which is described in more detail later in this book, the amount proposed by the Soviet government to deal with that accident in the national budget of the following year is said to have totalled roughly eight billion roubles.[1] According to an estimate by the Economic Research Institute of the National Academy of Sciences of Belarus, the economic losses in Belarus from 1986 to 2015 as a result of this accident will total 235 billion dollars (exchange rates in 1992). This amount is roughly equivalent to 32 times Belarus's national budget in 1985, and 21 times that country's national budget in 1991.[2] It has also been said that, if the Soviet Union had paid compensation for the economic loss suffered by other countries as a result of the accident, the Soviet economy would probably have collapsed.[3] This example serves to indicate the seriousness of the aftermath of an accident in a large-scale complex system.

The following examples illustrate the special features of accidents in complex large-scale systems.

Deep into the night of 2 December 1984, an explosion occurred at the Union Carbide India pesticide plant in Bhopal, India. That disaster can be regarded as a typical example of the extent of damage that can arise from an accident brought about by progress in science. The explosion caused the release of the toxic gas methyl isocyanic acid, killing over 2000 people in one night (3500, by one account).[4] Most of the victims were not employees of the factory, but rather people who lived in the vicinity. Counting only those affected directly, this accident affected nearly 50 000 people,[5] or about 6.25 per cent of the entire Bhopal population of 800 000. How high these numbers would leap if the indirect effects were to be included cannot even be imagined. Furthermore, as the immediate effects from this accident on the human body, blindness and damage to the respiratory system were conspicuous. Quite some time after the accident, however, there were frequent occurrences in the victims of illnesses of the circulatory system and the digestive system and of memory disorders, which were apparently caused by large amounts of chemical substances remaining in their bodies.[6] Beyond such illnesses, there have been reports of secondary diseases, such as births of deformed babies that are regarded to be caused by hormone abnormalities produced by thiocyanates, as well as suicides and criminal behaviour due to nervous disorders.

In the region around the plant, 1047 water buffalo died. Other animal deaths included many domestic animals such as cattle, goats, dogs, and horses, as animal corpses littered the streets.[7] Moreover, it has been reported that the number of animals that suffered respiratory difficulties and other such injuries rose to about 7000.

The accident investigation concluded that the cause was largely human error. The accident arose in part as the result of work performed in violation of rules, distrust of instrumentation, and neglect of emergency measures.

Workers first discovered a problem with the equipment at about 11:30 p.m. on the day of the accident. However, the local person in charge who received the report did not fully recognize the seriousness of the situation and responded by saying 'It's tea time. I'll have a look afterwards'.[8] This easy-going reaction greatly delayed the response to the situation and the consequence was a large explosion.

The problem was reported to the person in charge at about 11:45 p.m. The tea break ended at 0:40 a.m. the next day. Placing this 1 h break above inspection of the facility was one of the factors that brought about an accident of this proportion and resulted in such enormous loss. This is one example of how, even in a facility equipped with excellent technology, safety can easily be subverted by human operators through neglect or a low awareness of safety.

Incidentally, although the company that operated this plant was a multinational corporation with its head office in Danbury, Connecticut, the plant was closed after the accident, and the employees dismissed. In March 1986, a settlement was reached in the US courts by which Union Carbide would pay restitution amounting to 350 million dollars, but the Indian government demanded an additional 1 billion dollars in damages, and forbade Union Carbide from leaving Bhopal. Union Carbide's stocks fell rapidly after the accident, and it has been said that the company was forced to pay 4.8 billion dollars to prevent a takeover.[9]

The size of this settlement shows that, while the injury to human and animal life was enormous, the economic loss was also enormous.

1.2 The limits of safety measures based on accident investigations

Next, let us look at the accident at the Chernobyl nuclear power plant in the Soviet Union, which was literally a world-shaking event. A striking feature of this accident, which occurred on 26 April 1986, was that Sweden was the first to infer that the accident had happened. The radioactivity monitoring equipment at the Forsmark nuclear power plant in Sweden detected abnormally high levels of radiation and ascertained from meteorological conditions at the time that the source was in the Soviet Union.[10]

At that time, the weather in both Sweden and Finland was rain and light spring snowfall. The radioactive cloud from the Chernobyl accident was carried on a southeasterly wind to the coastline in southern Sweden in the afternoon of 27 April. In the Soviet Union, on the other hand, Secretary Gorbachev did not receive a detailed report on the accident until 28 April, 2 days after the accident.[11] Thus, the northern European countries were aware of the major accident 1 day before the Soviet Union itself, which bore the highest responsibility for the accident. This fact demonstrates the inadequacy and lack of transparency of the flow of information in the Soviet Union at that time.

In Japan, too, there have been information flow problems, such as the concealment of video recordings of the liquid sodium leak at the Monju fast breeder reactor that occurred in 1995,[12] the delay in reporting the accident that involved a radiation leak at the Fugen reactor in 1997,[13] and the cover-up of previous radiation leaks. Another was the fictitious report concerning a fire that occurred in the asphalt solidification facility in the Tokaimura nuclear fuel processing plant in Ibaragi Prefecture, which stated that the fire was under control (although there was only one confirmation, two confirmations were reported and an explosion occurred 10 h later).[14] It cannot be denied that, generally, in both eastern and western countries, as the danger when an accident occurs increases, so too does the tendency towards false reporting and concealing relevant information.

A number of similar accidents have occurred in the Soviet Union; notably, a major accident involving a nuclear explosion in 1957[15] and an accident involving the No. 3 reactor of the Beloyarsk nuclear power plant in 1978.[16] Concerning the Chernobyl plant, too, there had been various indications of safety-related problems, as suggested by the Russian journalist Lyubov Kovalevska.[17] In the court decision, presiding judge Raimond Brize pointed out that violations of safety regulations and negligence had become the normal state of affairs at the Chernobyl plant.[18] Prior to the accident, several incidents (problems that lead up to an accident; explained in detail in end note 52) had been experienced. If this kind of accident or incident information had been correctly understood, or if appropriate measures to improve the situation had been taken in response to these accusations, would it have been possible to prevent the Chernobyl accident? Or, in the event that the accident did still occur, would it have been possible to hold the damage to a minimum? Because the highest levels of the Soviet government did not receive the report of the accident until 2 days after the fact, later even than did other organizations of other countries, it is difficult to believe that the serious incidents that led up to the accident had been accurately and quickly reported to the offices responsible for studying improvement measures.

The effects of nuclear power plant accidents have a scope that extends beyond the immediate area around the accident site, and such accidents have the potential to affect a far broader area than did the Bhopal explosion. Moreover, the area that will be affected is subject to the meteorological conditions that follow the accident, and is thus impossible to predict. The radiation cloud from a nuclear accident moves unhindered by national borders, carrying radioactive material into neighbouring countries.

It was Poland that suffered the most damage from the Chernobyl accident. Because of the southeasterly wind,[19] Poland was contaminated across its entire expanse by radioactive material that was several hundred times as strong as that which affected the areas surrounding the Chernobyl plant itself.[20] Furthermore, the contamination was in the form of radioactive material fall-out in spots (microscopic particles of ruthenium), a kind of radioactive contamination that had not been measured before in nuclear experiments or other such situations, and entirely new types of damage have been reported.

Moreover, it has been reported that, even in Wales, some 1200 miles away from Chernobyl, 700 000 t of lamb and mutton (valued at about 20 million US dollars) was found to be contaminated by radiation and could not be shipped. The range of the effects of this accident is simply terrifying. It has been said that the Chernobyl plant must remain sealed for the next 200 years. It has also been said that some 100 000 people have been exposed to the danger of contracting cancer by this accident.[21]

To be sure, the losses due to nuclear accidents are beyond what can be imagined, and the scope of the effects and the forms of injury and damage are entirely new types that have not previously been experienced by humankind. In the face of this new type of danger, we must not fail to recognize the limitations of conventional retrospective safety measures that have been based on accident investigations.

1.3 The social impact of the Osutaka Mountain JAL plane crash accident

Accidents that involve large commercial aircraft, which are rightfully called complex large-scale systems, also have the characteristic features of very large systems. Although there have been quite a number of major airplane accidents, one that undeniably remains in memory is the crash of the Japan Airlines airliner on Osutaka Mountain, which was the world's largest single-plane accident. On 12 August 1985, the Japan Airlines flight 123 Boeing 747SR crashed into a steep mountain ridge near Osutaka Mountain. A total of 520 people died in this accident and four persons were miraculously rescued, although they suffered severe injuries.[22]

The Boeing 747 series is a 'third-generation' aircraft constructed on the basis of a fail-safe (redundant safety) design concept. This means that the system is configured such that, even if some part fails, that failure is limited to a part of the system and does not develop into a major failure or failures in other functions that might lead to an accident. In other words, the Boeing 747SR was said to have been the safest aircraft of its time. This accident was a real demonstration of the great extent and seriousness of the loss that can result from a single accident involving this type of aircraft, and the eyes of the world focused on it. Of course, the loss and repercussions of this accident extended beyond the passengers and crew, and those are explained in detail below, but first let us draw a brief outline of this tragic accident.

The number of people who died or were seriously injured in this accident is as mentioned above, but there were also three related suicides and one death in the line of duty within 3 months of the accident. In addition to the personal loss to the victims and their families, 18 companies lost as many as ten or more middle managerial personnel; 28 of the fatalities were company presidents or chief executives, so the effect on those companies was also severe.[23]

Medium and small enterprises often stand on the personal reputation of the company president or chief executive. Because of this accident, two companies were driven to virtual bankruptcy owing to the deaths of the company presidents; in another company, stop-gap assistance was given by a friend of the manager, and a music firm sank into functional paralysis because of the death of a popular singer. Such spreading influences of accidents have arisen as a new social problem.[24]

Next, let's look at the human labour and economic expense associated with the rescue operations and the investigation of this accident. To begin with, 30 volunteer groups participated in Fujioka City, where a safe repository for the corpses was set up, and a total of 3700 volunteers co-operated in the rescue and investigation operations, including the town of Uenomura in Gunma Prefecture, the site of the crash. Up to 386 people participated each day.[25]

The doctors involved in identifying the remains of the deceased, including dentists, numbered 1830. Because the bodies were so severely disfigured, no more than 90 of them could be identified by face alone. A total of 181 could be identified by clothing or personal belongings. The remaining 40 per cent or so were identified by dental records, which involved the co-operation of a total of 770 dentists.[26]

In addition, both the personnel of the local governments of Uenomura, Fujioka City and Gunma Prefecture, the Gunma prefectural police, and the Japan Self-Defence Forces that were dispatched for the rescue operation, as well as the associated expense, were of a scale beyond comparison with those of previous aircraft accidents. The Gunma prefectural police mobilized a total of 38 000 police officers for the search and rescue operations in August of that year alone. The incurred expenses totalled 427 million yen (about 3.5 million US dollars), including 256 million yen for overtime pay, 73 million yen for meals, 35 million yen for helicopter and vehicle fuel, and so on. Moreover, it has been reported that disbursements of 2–3 million yen (16 000–25 000 US dollars) continued day after day through September and afterwards.[27]

Ultimately, the expenses incurred by the three local government bodies amounted to about 800 million yen (about 6.7 million US dollars), 500 million yen (about 4.2 million US dollars) of which was reported by newspapers to have been billed to Japan Airlines, the company responsible for the accident.[28] The work following air accidents prior to this one, such as the Shizukuishi accident (on 30 July 1971), a midair collision between the commercial airlines aircraft Boeing 727-281 and the Self-Defence Forces training fighter North American F-86F, and the Tokyo Bay crash (9 February 1982), caused by pilot illness, was completed within a few days.[29] The direct expenses incurred for dealing with the accidents at the site were of the order of a few million yen according to newspaper reports. The scale of loss due to the Japan Airlines Osutaka Mountain crash could therefore also be said to stand out in terms of the expenses incurred. Moreover, the costs involved in investigating the cause of the accident, public and private, were also huge.

Of the 150 t of aircraft wreckage that was scattered around the crash site, a total of 44 t or so, including the 4 t engines and the 72 ft wide horizontal tailplane, that was required for the investigation had to be transported to the Ministry of Transportation hangars at Haneda.[30] That alone cost at least 100 million yen (about 800 000 US dollars). In addition to that, disposal of the parts of the wreckage that were not required for the accident investigation, the search of Sagami Bay, the investigation and analysis expenses of the Accident Investigation Committee, and other such costs probably totalled up to a considerable amount. We can also imagine that both Japan Airlines and the Boeing Corporation were charged with a very large amount, although those figures have not been made public. In spite of the payment of such huge amounts of money, investigation of the causes of the accident is not sufficient, as is explained later.

Another inescapable result of the occurrence of this accident is an increase in the cost of insurance, an effect that was not limited to Japan Airlines, but extended to other airlines as well. On 13 August, 1 day after the accident, the world's largest non-life insurer, Lloyd's, made an announcement suggesting that the insurance fees paid by international airlines companies would increase as a result of this accident.[31] Another negative effect of this accident was a decrease in the number of passengers. For the 1 month following the accident, from 13 August to 12 September, the number of passengers flying with Japan Airlines dropped to 87.7 per cent compared with the same period of the previous year.[32]

This outline of the effects and damages brought about by the crash of the Japan Airlines plane at Osutaka Mountain shows that it was more than an accident affecting one airline company, it was of a scale that should be expressed as a problem on the level of society and the entire public community. Accordingly, it is reasonable to say that the prevention of such accidents is not simply an issue for the affected airline company alone, but is rather a social issue.

1.4 Overconfidence in safety measures as a cause of accidents

Next, let's look at just what it was that caused this disaster.

In their report of this accident, the Aircraft Accident Investigation Commission of the Ministry of Transport Japan described the accident in the following way:[33]

The probable cause of this accident was the sudden rupture of the afterpressure bulkhead, causing damage to the tail fuselage, vertical stabilizer, and flight control system, which in turn caused impairment of the aircraft control and loss of the main aircraft control function. The cause of rupture of the afterpressure bulkhead was that, during the flight, it was so weakened by the fatigue cracks that had developed along the web connecting sections (the seam between the small plates of the pressure bulkhead) that the pressure bulkhead did not withstand the pressurization of the passenger compartment during flight.

The origination and development of the fatigue cracks resulted from improper repair of the same pressure bulkhead in 1978. It is inferred that the failure to discover the cracks during inspection and maintenance also contributed to the fact that the defect developed to the point of breaking.

These investigation results have received criticism from aviation experts, who claim that, although there was a gradual decrease in pressure, there was no confirmation[34] of the rapid decompression due to rupture of the bulkhead that would be sufficient to destroy the tail section fuselage, the vertical stabilizer, etc.; that is to say, the rapid decompression reaching

280 000 ft/min that is written in the accident investigation report. This represents decompression at the same rate as reduction to atmospheric pressure at 279 000 ft in the upper atmosphere in 1 min, or, expressed in another way, about the same as reduction to atmospheric pressure at the summit of Mt Everest at 27 880 ft in 6 s. The Accident Investigation Commission calculated that the air temperature dropped 65 K at once and to 40° C below freezing point as a result of this rapid pressure drop.[35]

If, however, the accident was caused by a decrease in operability of the aircraft owing to rupture of the bulkhead, as tentatively indicated by the Commission's report, first of all there is the matter of the improper repair of the rear bulkhead at the time of the accident in which the tail of this same aircraft struck the runway of the Osaka Airport in 1978.[36] The repair of the rear bulkhead was performed by the Boeing Corporation. In that repair, in part of a seam that should have been fastened with two rows of rivets, only a single row of rivets had been used. As a result, the strength of that part was reduced to about 70 per cent of the original strength, and it was inferred that fatigue cracks could easily occur under that condition.[37]

This improper repair was not discovered by the Boeing repair inspectors, was not discovered by the Japan Airlines inspectors in the post-repair acceptance inspection, and also was not discovered in the inspection by the Civil Aviation Bureau of the Ministry of Transport. Moreover, the improper repair was not discovered in as many as six maintenance inspections that were performed after every 3000 h of aircraft operation, nor were the cracks that were supposed to have already occurred in the repaired part discovered in those maintenance inspections. The Boeing 747 series plane that was involved in the accident was constructed on the fail-safe design concept and was considered to be the safest aircraft of its time. It is possible that this led many people to become overconfident in the safety of the airframe of this aircraft. It is reasonable to assume that such overconfidence could lie behind the improper repair and the subsequent failure to see that improper repair. Concerning the Chernobyl accident mentioned earlier, Hans Blix, former Director General of the International Atomic Energy Agency (IAEA), has stated the opinion that safety is the result of the persons in charge being aware of potential crises, while danger is high when the persons in charge make light of crises from the beginning, with the mind-set that safety exists. The report of the President's Commission on the accident at Three Mile Island, 28 March 1979, which investigated the accident at the Three Mile Island nuclear power station in the United States,[38] stated that 'The biggest mistaken conviction is that all of the station personnel trusted in the safety of the facility, and that there was an insufficient concern for human factors in atomic power generation'. These words not only apply to safety in atomic power generation but have much to suggest concerning safety in any field.

1.5 Humility with respect to the facts

What is most important in terms of basic attitude towards safety is humility in the face of facts. That is to say, seeing facts directly, without preconceptions and without taking a particular standpoint. This is the attitude that is necessary when investigating facts, regardless of the field, and, although it is a matter of course, this is an extremely important attitude for the investigation of accidents as well as incidents.

In order to gain prestige and wrest away leadership in air transportation over the Atlantic Ocean from the United States after World War II, the British developed an aircraft named the 'Comet'. After commissioning of the Comet in 1952, however, four accidents occurred, and, although accident investigations were conducted, no clear causes were identified.

Then, on 10 January 1954, an accident occurred in which a Comet that had taken off from Rome disintegrated during climb over Elba Island in the Mediterranean Sea.[39] The British Overseas Airways Corporation (BOAC) halted flights on its own initiative, considered over 50 items and formulated countermeasures for all of them (for example, inserting reinforcement plates so that, even if an engine exploded, the fuselage would not be damaged), and then resumed flights two and a half months later.

Two weeks after that, however, on 8 April, another accident occurred in which a Comet aircraft that had taken off from Rome disintegrated during flight at high altitude (30 000 ft) and crashed near the island of Stromboli in Italy.[40] The very same day, the British Civil Aviation Bureau revoked the aircraft type certificate for the Comet (which certifies that the design for an aircraft series conforms to safety standards) and Prime Minister Churchill ordered a thorough investigation.[41] The revoking of the Comet's aircraft type license resulted in the loss of a 40 million pound (about 64 million US dollars) contract for the aircraft manufacturer DeHaviland.

First, pressurization tests were performed using the actual aircraft. The results showed that cracks developed along the rivets around the window frames and automatic direction finder (ADF) antenna.[42] Furthermore, to recover the fuselage from the Elba Island accident, a precise scale model was exploded and the scattering of the debris was studied. Because the recovery of the wreckage from the deep sea around the Stromboli Islands was technologically impossible at that time, the investigation focused on the Elba Island accident, which was extremely similar to the 8 April accident. On the basis of the experimental results, and from the wreckage recovered from the sea, it was confirmed that cracking proceeded owing to repetitive stress on the frame corners and the ADF antenna area, causing disintegration in flight.

Because of the Comet accidents, Britain was not able to wrest leadership in civil air transport over the Atlantic away from the United States. Nevertheless, the basic posture of a persistent investigation based on facts and the great effort expended on that investigation on the order of the Prime Minister are worthy of praise. The results of this investigation contributed greatly to safety in subsequent civilian air transportation. For example, after these results were known, the corners of all openings in aircraft were rounded so that cracks could not occur, and that became basic in aircraft design.

This sort of sincere attitude towards accident investigation has continued on in the aircraft manufacturing countries, notably Britain and the United States.

After the crash of the South African Airways flight 295 Boeing 747-combi into the sea about 150 miles north-east of Mauritius on 28 November 1987, the joint investigation commission formed by the countries involved calculated the location of the crash site from the content of the communications of the plane with air traffic controllers and recovered the cockpit voice recorder (CVR) from the seabed at a depth of 14 760 ft.[43]

In the crash (due to an explosion) of the Air India flight 182 Boeing 747-237B in the sea south-west of Ireland on 23 June 1985, the aircraft experienced a rapid decompression similar to that of the aircraft in the Japan Airlines Osutaka Mountain accident. In the investigation following the crash, many fragments of the aircraft were also recovered, including the FDR, CVR, and engine nacelle (the outer covering that houses the engine) from the sea floor at 7200 ft below sea level.[44]

Furthermore, after the 24 February 1989 accident involving United Airlines flight 811 Boeing 747 out of Honolulu, in which the door of the lower freight compartment opened and nine passengers were sucked out of the aircraft as the plane was climbing and approached an altitude of 24 000 ft, the door that fell off was recovered from the ocean. Although initially it had been suspected that the door might not have been closed completely,[45] examination of the recovered door revealed that the door lock had opened as the result of a short circuit in the electrical system.[46]

1.6 The unrecovered vertical tailplane fragment of the Japan Airlines aircraft

Concerning the accident involving Japan Airlines flight 123 at Osutaka Mountain, although most of the fragments of the vertical tailplane, which was directly related to the occurrence of the accident, were inferred to have been scattered around the Izu Peninsula, those fragments have not been completely recovered.

The Aircraft Accident Investigation Commission of Japan wanted to substantiate that the outer plates of the vertical tailplane were peeled off by air pressure after the rivet heads had come out and the plates had lifted upwards and off as a result of the rapid decompression caused by the rupturing of the pressure bulkhead. It thus constructed a full-size partial model of the vertical stabilizer torque box – a support member that has a box-like structure whose purpose is to maintain the strength of the vertical stabilizer – and conducted tests in which pressure was increased until the model ruptured. The test results showed that the upper part of the torque box ruptured when the pressure within the torque box reached about $4 \, lbf/in^2$.[47]

The Commission's report stated that destruction of one part spread successively to adjacent parts, and, as that destruction proceeded, the outer plate bulged out under the effect of the large difference between internal and external pressures and was peeled off by airflow.

The results of that experiment, however, are the same as the results of experiments performed using the actual tail section of the British-European Airlines (BEA) Vickers Vanguard aircraft flight 706 that crashed over Belgium in 1971, 14 years before the Osutaka Mountain accident. In the BEA accident, the aircraft made a virtually vertical descent after destruction of the horizontal tailplanes following a rapid decompression. The accident report written by the joint Belgian and British investigation committee that investigated the Vickers Vanguard accident[48] contained the following comments with regard to the pressure tests on the tail section:

The definitive test involved increasing the cabin differential pressure to $6.25 \, lbf/in^2$ before opening the petal valve to simulate a rear pressure bulkhead failure at an altitude of $19\,000 \, ft$. The rear tailcone reached a maximum of $5.5 \, lbf/in^2$ after $0.12 \, s$. Some $0.03 \, s$ later there was a sudden *pressure fluctuation in the right outboard tailplane when the pressure there was 4.12 lbf/in².*

This test produced severe distortion of the right-hand tailplane upper surface skins between tailplane stations 143 and 276. The worst damage was along the chord at the elevator hinge rib at station 213 where rivet heads had pulled through the skin. On the right-hand side of the fin the skin was distorted between the centre and rear spars. Rivet heads had pulled through the skin at fin stations 117.8 and 154.1.

Internally, there were extensive failures of cleat to stringer rivets in the right tailplane.

A study of the film record in conjunction with the pressure traces showed that the disruption of the tailplane started in the region of station 234, spreading outboard and inboard.

In comparison, according to the records of the Osutaka Mountain accident investigation report (test research materials related to JA8119), the results of the internal structural destruction tests for the vertical tail fin showed burning sounds at $3.3 \, \text{lbf/in}^2$, the sound of metal expanding at $4.25 \, \text{lbf/in}^2$, and burning sounds at $4.28 \, \text{lbf/in}^2$, and showed that a pressure increase was impossible. An examination of the destruction condition of the test sample showed that the attachment segment of the stringer and rip cord across nearly the entire right side had been destroyed, that 80 per cent of the rivets securing the right-side external plates and the shear ties had come loose, and that considerable residual distortion could be seen in the right-side external plates.[49]

In other words, on the Vickers Vanguard aircraft, there was a sudden pressure fluctuation when the pressure inside the horizontal tail fin reached $4.12 \, \text{lbf/in}^2$, the rivet heads were left behind, and the external plate lifted up. In the Osutaka Mountain accident, a test of part of the vertical tail fin torque box showed that many of the rivets were destroyed when the internal pressure reached $4.28 \, \text{lbf/in}^2$, and that a pressure increase was impossible.

The Vickers Vanguard experiments included tests with the same rivets as used in the plane that crashed and tests with improved, strengthened rivets, but, in the JAL case, tests with improved rivets were not performed.

The joint Belgian and British investigation committee that investigated the Vickers Vanguard accident indicated that, from the results of the experiment, it was necessary to improve the tail section of the aircraft to protect against the leakage of a large volume of pressurized air from the pressurized compartment. The use of strengthened rivets had been demonstrated to prevent such destruction and it was indicated that consideration should be given to whether or not certification of airworthiness should be issued to aircraft that do not use strengthened rivets.

Specifically, the report[50] states:

However, it may be significant that following the tests carried out in the course of the investigation, it was found that the left tailplane of the specimen, which was modified to the later standard, showed no signs of distress or incipient failure. Considerations should be given by the airworthiness authorities concerned as to whether tailplanes that do not incorporate modification L40 should be kept in service.

This was 2 years before the Boeing 747SR series aircraft that was involved in the Osutaka Mountain accident was put into service in Japan. If we take into consideration both the comments in the above accident report and the time that this model was introduced into Japan, we can assume that, when the Boeing 747SR was introduced into Japan, the Minister of Transport should have taken the above comments into account, confirmed and studied whether the construction surrounding the stabilizer could withstand a pressure leak in the event that the cabin wall ruptured and an air pressure leak occurred, and ultimately should not have certified the aircraft as being airworthy. In other words, perhaps there was a problem in government policy regarding aviation when the model was introduced. Furthermore, perhaps the fact that there was no system for utilizing the lessons of the Vickers Vanguard accident indicated a problem not only with aviation administrative policy in Japan but also with the Boeing Corporation, which designed and built the aircraft. Moreover, after this accident, did the Minister of Transport conduct investigations when the A320 (1991), MD-11 (1994), and B-777 (1995) series aircraft were introduced, to determine whether the rivets had been strengthened sufficiently? It would certainly not be unreasonable to harbour such doubts as these. Could it not be said that the investigation of the accident was insufficient in that recovery of the vertical tailplane fragments that had sunk in the sea around the Izu Peninsula was necessary in order to determine the cause of the accident?

The author is probably not alone in entertaining such questions. In light of such matters, even the so-called 'tombstone safety' is questionable. Tombstone safety refers to increased safety that results from improvements made on the basis of investigations of fatal accidents.

The attitude of humility in the face of facts is also a required posture in the face of accidents that involve human casualties. Accident investigations are not simply the identification of the cause of an accident in terms of the hardware of the aircraft; they must also serve to improve the safety and survivability of human life when an accident occurs. With respect to both the Civil Aeronautics Law of Japan and the Convention on International Civil Aviation – the basic convention on international civil aviation, with 188 signatory states at the time of writing in 2003, including Japan – the investigations of the Aircraft Accident Investigation Commission of Japan concerning the identification of the causes of the human deaths and injuries have been insufficient, in spite of the fact that human deaths and injuries in which aircraft played a part are considered aircraft accidents. This is not only true in the case of the Osutaka Mountain accident.

Injuries to the human body by aircraft include injuries to the eardrum by rapid decompression, burns from fire, suffocation due to inhalation of toxic gas released from cabinet trim and so on, as well as broken bones, contusions, lacerations, and internal organ injuries suffered in crashes. Analysis of blood for the levels of toxic gases present or examination of the lung tissue of victims and other such post-mortem examinations are important in determining the cause of death.

Concerning the Japan Airlines Osutaka Mountain accident, it is widely known that, aside from the four persons who were rescued, there were others who remained alive for some time after the crash. Nevertheless, the investigation report simply stated that the passengers 'either died instantly or nearly so'; there was no detailed verification of the causes of death.

In contrast, the British Aircraft Accident Investigation Board performed tissue examinations on 26 of the 131 bodies recovered from the Air India flight 182 Boeing 747 that experienced the same sort of rapid decompression as the Japan Airlines plane and crashed into the sea near Ireland on 23 June 1985. The Board also performed examinations for eardrum damage in 25 of these bodies, including children who had not suffered from skull damage, to provide evidence of the rapid decompression.[51] To ensure human safety, the most important matter in air transportation, should the Aircraft Accident Investigation Commission not have a role in making improvements for increasing survivability by conducting detailed investigations, seeking the facts with greater humility?

1.7 Incidents and the structure of an accident

As mentioned earlier, prior to the Chernobyl nuclear power plant accident, a number of problems that could be referred to as incidents occurred. It is entirely natural to imagine that, if suitable measures had been taken at the stage of such incidents, they might not have developed into an accident.

On the basis of the results of a scientific survey of some 7500 industrial accidents, H. W. Heinrich once succeeded in deriving the probability distributions of safety-related events, referred to as Heinrich's law. It is expressed as the 300–29–1 ratio opportunity, which means that, prior to the occurrence of one major accident, there occur 29 minor accidents and 300 problems that do not involve damage (i.e. do not reach the point of becoming accidents). That is to say, before a major accident occurs, there is a frequent occurrence of what could be called 'incidents' (see Fig. 5).

Whether or not this Heinrich 300–29–1 ratio actually applies to technologically advanced complex large-scale systems, it is not hard to imagine that the occurrence of major accidents is preceded by many

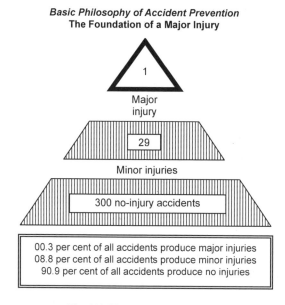

Basic Philosophy of Accident Prevention
The Foundation of a Major Injury

1

Major injury

29

Minor injuries

300 no-injury accidents

00.3 per cent of all accidents produce major injuries
08.8 per cent of all accidents produce minor injuries
90.9 per cent of all accidents produce no injuries

The 300-29-1 ratio spells opportunity.
H.W. Heinrich: Industrial Accident Prevention
— A Scientific Approach (4th ed. P.27)

Fig. 5 Heinrich's law

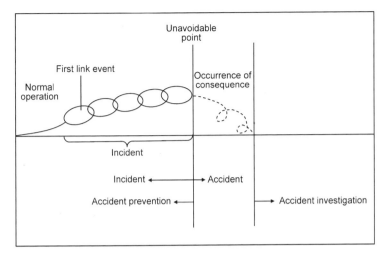

Fig. 6 Unsafe events (chain of events)

small accidents and the experiencing of problems that might be called 'frights', 'starts', or 'close calls'.

Here, to clarify the relationship between accidents and incidents, let us consider the process by which an accident occurs. In the process of an accident, multiple danger factors, generally of different dimensions, are intertwined with human errors and linked together (a chain of events). When the development towards an accident reaches an unavoidable point, an explosion, crash, collision, or other such deterministic situation (accident) is considered to occur (see Fig. 6).

Even in such cases, however, it is possible to return to a state of normal operation if corrective operations are performed after operators are made aware of abnormal circumstances as the result of the person noticing by himself, receiving advice or indication from another person, or a warning device. A case in which it is possible to return to a state of normal operation is referred to as an 'incident' (pre-accident event) as opposed to an 'accident'. The term 'incident' has various meanings in different fields, so it is necessary firstly to clarify the concept of incident as we use it here. The incidents that we take up in this book from the viewpoint of accident prevention and the securing of safety, regardless of the field of interest, are 'unsafe events other than accidents that have affected safety or may have affected safety'.

That is to say, the incidents in the field of aviation that are referred to in this book are not limited to the 'serious incidents' of Annex 13 of the Convention on International Civil Aviation, but include events other than accidents that have affected safety or may have affected safety in relation to the operation of aircraft. In the field of nuclear power, the incidents are not limited to those that pertain to levels 1 to 3 on the scale for evaluating the severity of international nuclear incidents, but include those incidents of level 0 and even those that are not subject to evaluation but broadly affect safety.[52] The concept of 'unsafe event' includes incidents as well as accidents. The danger factors that evolve into an incident also have the potential to develop into an accident because, if the chain of danger factors had not been broken, an accident would have occurred. Possible reasons for restoration of normal operation might be an act of confirmation by an operator, advice or indication from another person, or the functioning of a warning device.

Accordingly, if it were possible to use the data on incidents to identify the danger factors that have the potential to lead up to an accident, and, by analysing and studying incidents, to determine the relationship between human errors and the danger factors to which human error is susceptible, and ascertain why incidents did not proceed to develop into accidents (reason for recovery) and the degree of the danger posed by them, it would become possible to consider appropriate measures for preventing accidents. That is to say, if accidents could be foreseen or predicted and appropriate measures taken on the basis of latent danger factors that have been identified from large amounts of data on the minor accidents and problems that constitute the lower position in Heinrich's law, then we could derive a logic for preventing major accidents. The identification of incidents that involve the potential danger factors that constitute the accident process is thus critical to the prevention of major accidents.

When thinking about the securing of safety, no matter what the field, phenomena that affect safety or phenomena that could potentially affect safety have an extremely important meaning regardless of whether or not they fall under the definition of a 'serious incident'. This is particularly true in the case of atomic power, regarding safety phenomena that were not related to the release of radioactive materials or to defence in depth.

1.8 Errors made by a veteran pilot

The collision of the KLM (Royal Dutch Airlines) flight 4805 and PAA (Pan American Airlines) flight 1736 jumbo jet airliners on the runway of Tenerife Airport in the Spanish Canary Islands on 27 March 1977 is the world's largest accident in civil aviation.[53] To clarify the structure of an accident, I will use the process of that accident as an example in an attempt to provide a concrete explanation of whether or not there are opportunities to check the process of development into an accident in the chain of danger factors that constitute an accident, or the process of the chain of events. I will also use this example specifically to explain some of the background factors that lead to the errors made by humans.

The investigation of the Tenerife accident was conducted by the government of Spain, the state of occurrence, in accordance with the Convention on International Civil Aviation. In major international accidents, however, the parties involved have differing interests and positions concerning the accident, and there are a variety of interpretations and opinions based on various standpoints. Here, with these standpoints in mind, I will attempt an explanation based on the results of an analysis of the CVRs of both aircraft described in an accident report by the Spanish government released in ICAO's Accident Digest, as transcribed by the National Transportation Safety Board (NTSB).

The chain of events began with the fact that both of the planes had to be diverted to Tenerife as the Las Palmas Airport, their original destination, was closed because a terrorist group had exploded a bomb in the terminal building. Several other aircraft had also been diverted to Tenerife, which is a local airport with a small apron and a single runway. The airport was thus congested to the extent that, even if the Las Palmas Airport reopened, the departing planes could not easily exit to the runway (see Fig. 7).

The control tower had instructed the KLM aircraft to proceed to the end of the runway and make a 180° turn, because the plane was facing in the direction opposite the take-off direction (i.e. north-west to south-east).

When the KLM plane had completed its turn, arrived at the take-off stand-by position, and begun preparation for take-off, the PAA plane, which had received permission from the control tower to follow the KLM plane in the south-eastern direction on the same runway, had passed through the second taxiway and was proceeding towards the turned KLM plane, which was facing it on the same runway.

The control tower had planned to have the PAA leave the runway and proceed to the take-off stand-by position by way of the taxiway before the KLM plane was to take off.

There are four taxiways leading off the runway, C-1, C-2, C-3, and C-4, in the direction of the take-off point (i.e. to the southeast).

Taxiway C-1 intersects the runway at a 90° angle. Taxiways C-2 and C-3 are angled such that the aircraft would have to make a sharp left turn, and C-3 makes an angle of 146° with the runway. Taxiway C-4

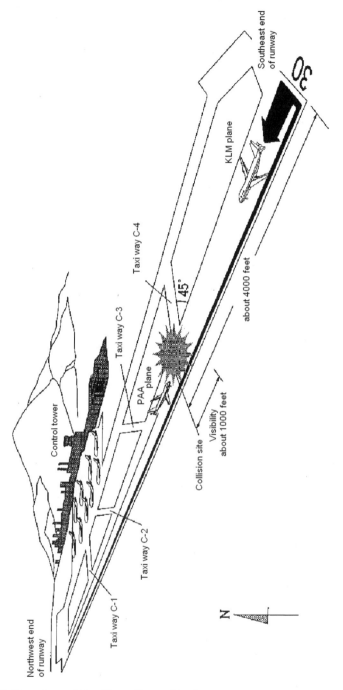

Fig. 7 Overview of the Tenerife accident
(modification of a diagram by Don Mackay in Time Magazine)

angles off at about 45°, and thus provides the best exit for the jumbo jet, allowing it to leave the runway with the easiest manoeuvre by completing a 45° turn. The taxiways are narrower than the runway, so, if a large aircraft makes a sharp turn of 146°, it is highly possible that the wheels will run off the runway, making it very difficult to manoeuvre the aircraft, so care is needed. The collision occurred after the PAA plane had passed taxiway C-3 and just before it had reached C-4.

The determining factor in the accident was that the KLM pilot proceeded to take off without receiving permission from the control tower. In international civil aviation, clearance from the control tower is required for take-off, and the term for that permission is 'Cleared for take-off', even in non-English-speaking areas. The aircraft that receives that permission is required to respond to the control tower by giving the call sign of the aircraft and repeating the instruction.

The captain of the KLM plane was a senior pilot at the training centre of that company and had 21 000 h of flight experience. One wonders why he violated this most basic rule by taking off without clearance from the control tower. This is a question that anyone would raise, all the more so because this was the most skilled and trusted pilot in the company. In the background, however, is the limit on working time imposed by the company, of which the captain must have been strongly aware.

Up to a few years before this accident, it was possible for a pilot to extend the working time of other flight crew members on his own decision in order to complete a task. Later, however, that pilot discretion was disallowed, and exceeding the working time limit was made subject to punishment, with the expectation that the limitation would be observed.

The formula for calculating the working time limit became very complex after 1977. Because it could not be calculated in the cockpit, the captain had to contact the airline's office in Amsterdam. The response was that, if it were not possible to take off before a certain time, there was a possibility that the working time limit would be exceeded.

That plane had to stop at Las Palmas to pick up waiting passengers and return to Amsterdam. Delays due to congestion at Las Palmas were expected, so, to avoid the delay of refuelling there, the captain had taken on sufficient fuel at Tenerife. Although there is testimony by maintenance personnel that the pilot's attitude prior to departure at Tenerife was calm and friendly, it can be inferred that anxiety over the problem of the severity of the working time limit and the inconvenience to the passengers of the stop at Las Palmas was affecting him strongly.

The captain had experienced far more problems than for a normal flight, including the change in landing site, the 180° turning manoeuvre on the end of a 150 ft wide runway, refuelling, and radio contact with the company office, and he was surely feeling fatigue. The same was true for the copilot. Also, although the captain had flown the air routes in Europe and between continents for many years, he had been serving as a senior pilot at the training centre for a period of more than 10 years, and so it has been inferred that his sensitivity to the actual passenger flight operations may have been reduced.

When training is conducted by flight simulator, the instructor plays the role of the air traffic controller and is in the position of granting permission to take off to his students. Usually, the flight control operation training is conducted without making a problem of permission from air traffic control.

1.9 Background to mistakes made by the air traffic controller

It is clear, however, that the error of the KLM pilot was not the only cause of the accident. Tenerife Airport is not only high above sea level but also situated in a depression and hemmed in by mountains. Low clouds that can change visibility form easily, and those typical weather conditions existed on the day of the accident. According to the official data recorded in the accident report, the runway visibility 17 min before the accident occurred was from 1.2 to 2 miles. Five minutes before the accident, however, the visibility had dropped to 1000 ft. This obstruction of visibility by ground-level clouds was also a background factor in the accident.

If the runway visibility had not been greatly reduced by the low clouds, even if the KLM plane had begun to take off without tower clearance, both of the planes would have been able to see the other plane approaching. The KLM plane would then have taken steps to reject the take-off, and the PAA plane, on the other hand, would have driven off the runway to avoid the collision.

The accident report by the Spanish government stated that the special weather conditions of Tenerife Airport should themselves be considered a factor in the accident. The problem, however, is that the air traffic controller who was controlling the airport traffic, and who should have set the distance between the aircraft visually, did not use the ground radar and failed sufficiently to confirm the current positions of the two planes through verbal exchange. Under low visibility, the controller therefore did not have accurate knowledge of the relative positions of the two aircraft.

If the air traffic controller had known the position of the moving PAA plane and the take-off ready status of the KLM plane, he could have given strong and clear instruction to the KLM plane that it could not take off until the PAA plane had left the runway, and could probably have stopped the situation from developing into an accident. For an air traffic controller, having two aircraft facing each other on the same runway is the situation that must be avoided above all costs.

Under the minimum weather conditions for permitting take-off at Tenerife Airport, the visibility must be at least 2600 ft when the runway centre-line lights are not illuminated. It is clear that the pilots of both planes were fully aware of this restriction; because the KLM pilot had checked with the air traffic controller as to whether or not the centre-line lights could be used, and because the PAA crew had discussed the take-off restriction among themselves in the cockpit, and believed that the minimum weather conditions were not satisfied at that time. Nevertheless, there was no report from the air traffic controller of either the latest weather information just prior to take-off or the suddenly changing runway visibility, and there were no inquiries to confirm the conditions from either aircraft.

In the job of the airport controller, whose responsibility is to avert aircraft collisions by visual observation, if the controller cannot discern both planes either visually or by means of some aid to vision, it is improper to allow two aircraft to be on the same runway at the same time, even if it means temporarily closing the airport, as that is a major cause of accidents. Our point here is that the natural phenomenon of low clouds should be considered a background factor that led to the error of the air traffic controller, and not a direct cause of the accident.

1.10 Improper instructions by the air traffic controller

If we view the circumstances of the accident as a time series, we can see that the error of the KLM crew, the inappropriate instruction from the air traffic controller, and the failure of the PAA crew to confirm the ambiguous instruction in which the term 'third' was used, among other factors, are linked together in a chain.

The cockpit voice recorder of the KLM plane was manufactured in the United States. The voice recording contained much noise and reverberation. After eliminating some of that noise, the content of the recording was transcribed by the US National Transportation Safety

Board. According to that recording, the pilot of the KLM plane began the take-off operation with only what is called the climb instruction from the air traffic controller, and without clearance to take off.

After reading back the climb instruction to the air traffic controller, the copilot announced 'We are now at take-off'. To request clearance for take-off, the copilot should say 'We are ready for take-off' or 'We are standing by for take-off'. The use of a non-standard, unusual expression in communication with the controller not only made the crew's intention unclear but was also improper. The copilot should have stated clearly to the pilot that clearance to take off had not been given and stopped the pilot's take-off.

Although the copilot had had a total of 9200 flying hours, no more than 95 of those hours had been on the Boeing 747. He, too, was surely feeling fatigue, as was the captain, and it has been inferred that he may have hesitated to question or to voice a difference in opinion to the most respected captain in his company. Even when a person has doubts concerning the action of a more experienced, trusted superior, there is a tendency to think that there might be some reason unknown to that person, and this tendency stops the person from frankly questioning the action. This is a situation that has probably been experienced by many people in other work situations as well.

The air traffic controller either misheard the copilot's transmission of 'We are now at take-off', which was an improper expression but had an extremely important meaning, as 'We are now at take-off position', or did not understand the situation and simply answered 'OK' followed by 'Stand by for take-off. I will call your aircraft'. It has been pointed out that the term 'OK' used by the air traffic controller was also improper. The Dutch government pointed out that the air traffic controller should have listened for the pilot's confirmation of the controller's instruction to 'Stand by for take-off'.[54]

The air traffic controller, meanwhile, had instructed the PAA plane, which was moving from north-west to south-east on the same runway as the KLM plane, to exit the runway at the 'third' taxiway. Although each of the taxiways has a name that unmistakably identifies it, the controller simply used the expression 'third' rather than the specific taxiway name. The PAA crew that received the instruction to make the third left turn did not turn left at taxiway C-3, but proceeded directly towards C-4 without confirming with the control whether the instruction meant C-3 or C-4.

The conclusion of the accident report by the Spanish government indicated that the failure of the PAA aircraft to turn left to exit the runway at the third intersection, which is to say C-3, was also a factor that contributed to the accident. That report also pointed to the possibility that the PAA crew intentionally passed by C-3. However, if the controller had from the beginning intended the plane to leave the runway via C-3, the term 'third' was obviously not a clear and accurate indication for taxiway C-3. Furthermore, the instruction to have a jumbo jet make a sharp left turn (146°) off a narrow runway into a taxiway was, in itself, improper.

Moreover, from the conversation among the crew in the cockpit, it can be inferred that the PAA plane had planned from the beginning to exit the runway at C-4, where the turn manoeuvre was easy, but they did not confirm whether 'third' referred to C-3 or C-4. There was repeated confirmation with the traffic controller, however, regarding whether or not it was 'third'.

1.11 Unbroken chain of events leading to the accident

Although the Dutch government held that the fact that the PAA plane had not turned at taxiway C-3 and was still on the runway was a contributing cause of the accident, it can be considered that there is no cause-and-effect relationship between the PAA plane not turning and the collision.

If the PAA plane had attempted to turn left at C-3, a large-angle rotation of the aircraft on the ground would have been required, and the turn would have to have been negotiated with the fuselage of the plane near the right edge of the runway and after slowing down to a near stop. Accordingly, further time would be required for the manoeuvre and thus the collision might not have been prevented. Also, the traffic controller would probably not have considered giving clearance for the KLM plane to take off, because that PAA plane was still moving on the runway, and in fact that clearance had not been given. However, it is necessary for the aircraft crew to confirm instructions that use such unclear terms, and, in this respect, the accident leaves us with an important lesson.

After passing taxiway C-3, the PAA plane overheard the 'We are at take-off' transmission of the KLM plane to the control tower, and transmitted 'This is flight 1736. We are still on the runway'. That transmission was superimposed with the tower's transmission to the KLM plane of 'Stand by for take-off; we will call you'. There was thus 3 s of continuous shrill noise in the KLM plane owing to the two simultaneous transmissions, and the KLM crew was not able to understand the message.

The Spanish government also took the fact that these two transmissions were made simultaneously to mean that the transmissions were not received with the desired clarity, and pointed to this as one of the causes of the accident. If the KLM crew had been able to know that the PAA plane was still on the runway, there is no doubt that they would have taken steps to reject the take-off. Even at this point there was a missed opportunity to break the chain of events that led to the accident.

Upon receiving the message from the PAA plane that it was still on the runway, the traffic controller instructed 'Report runway clear', to which the PAA plane responded with 'OK, will report when we're clear'. The exchange between the PAA plane and the tower was also received by the KLM plane, which had begun its take-off run. The KLM flight engineer twice threw questions at the captain, asking whether the PAA plane was not still on the runway. The captain responded to the questions with 'Oh, yes (they have left the runway)' in a fully confident tone of voice and continued with the take-off operation.[55]

The captain, as the Dutch government has maintained, held a strong conviction that he had received clearance for take-off and that the PAA plane had left the runway. The captain therefore was probably not psychologically prepared to check with the control tower concerning the flight engineer's questions. Still, was there not at least a need for the copilot, who is responsible for communication with the tower, to confirm the location of the PAA plane? At that point in time, it should still have been possible for the KLM plane to reject take-off, even though 20 s had passed since the plane had begun acceleration for take-off. If steps had been taken to reject take-off at this point, it would probably have been possible to reduce the scope of the damage, even if it were not possible to avert the collision entirely.

Ten or more seconds later, the PAA crew suddenly saw the KLM plane charging towards their plane from the middle of a dense low-lying cloud and tried desperately to get their own plane off the runway to avert a collision. However, the landing gear of the KLM plane, which had lifted off the runway, directly struck the No. 3 engine, stripping off the upper part of the fuselage and cutting off the vertical tailplane. At about the same time, an explosion and fire occurred, resulting in major destruction.

After the collision, the KLM plane touched down again and came to a stop after sliding and spinning on the runway. The plane was enveloped in flames and not one of the 257 passengers and crew members was able to escape. As for the PAA plane, 326 of the 396 passengers died, including those that had been rescued but later died in hospital. Sixty-one persons, including two crew members, miraculously survived. A total of 583 persons died in this accident.

Here we have examined the Tenerife accident, but this tragedy is representative of other accidents in that they occur when danger factors on several dimensions intertwine with human errors in a complex way, and nothing acts to break the chain of events.

1.12 Lessons not applied

Next, to clarify the relationship between accident and incident, we will take another look at the danger factors that led to the Tenerife accident, and try to find past examples in which those factors were previously experienced in other places by other crews.

In the past, specific 'fright', 'start', or 'close call' experiences that correspond to the '300' level of Heinrich's law have, in the field of aviation too, been passed down by word of mouth in an extremely limited range, but there has been no systematic collection of such experiences recorded as examples. For this reason, it is not possible to present here a set of examples of the same danger factors as lie behind the Tenerife accident like the Heinrich '300'. Based on experiences related by many pilots, however, these same danger factors are experienced, and frequently, and there is no doubt that they occurred on the day of the Tenerife accident as well. Moreover, accidents that occurred before the Tenerife accident may have been caused by these same factors.

An accident occurred in Anchorage on 16 December 1975 in which the aircraft skidded on an icy taxiway on the way to the take-off position and slid down a slope, resulting in injuries to two persons. In that accident, the pilot was anxious to land at the destination airport within the take-off and landing time limits in the same way that the pilot of the KLM plane, being aware of that company's working time limits, was anxious to take off quickly. That accident involved Japan Airlines flight 422 Boeing 747 and occurred 1 year and 4 months before the Tenerife accident.

Just before departure, the cockpit crew of flight 422 had received a telex instructing that they could not land at Haneda International Airport after 23:00 Japan time because of the take-off and landing time limits. Based on that information, the crew calculated that the plane would have to take off by 21:00 local time at the latest. If the limit time was passed, it would be necessary to change the landing site and land at an alternative airport, creating an extreme inconvenience to the passengers.

The US National Transportation Safety Board, which investigated the accident, cited loss of directional control owing to the icy taxiway and strong crosswinds as the primary cause of the accident. The Board pointed to the delayed and insufficient measures taken by the airport manager against the ice on the taxiway, even though it was predictable, and to the pilot's improper decision to fly in spite of having received a report of the bad surface condition of the taxiway, placing priority on the schedule. It has been pointed out that it was wrong to give consideration to the schedule in the decision concerning bad operating conditions.[56]

Also, there was an accident at the Chicago O'Hare International Airport on 20 December 1972, 4 years and 3 months before the accident at Tenerife. In that accident, two aircraft collided on the ground while moving on the same runway under low visibility due to dense fog as the result of unclear communication with the traffic controller, the same situation as at Tenerife. The aircraft involved in the collision were the Delta Airlines flight 954 Convair 880, which was landing, and the North Central Airlines flight 575 DC-9, which was departing.

In addition to the two parallel runways [32-R (right) and 32-L (left)], O'Hare Airport has four intersecting runways, counting only the main ones. When giving instructions concerning the waiting circle to flight 954, which had landed and was moving on the taxiway, the air traffic controller specified simply '32', omitting the important R or L. When the pilot of flight 954 received that instruction, he had the wrong conviction that it meant 32L, considering the position of his plane. Without confirming the meaning, he cut across runway 27-L to proceed towards the 32-L waiting circle.

At that time, the DC-9, which was moving towards runway 27-L (which intersects runway 32-L) to take off, suddenly discovered the other aircraft on the runway. In spite of an attempted collision avoidance manoeuvre, flight 575 collided with the vertical stabilizer of the flight 954 Convair 880. The flight 575 DC-9 caught fire, and, as the escape slides did not open, ten persons died, including nine passengers whose escape was delayed. However, everyone on the flight 954 Convair 880 escaped.

As the presumed causes of this accident, the US National Transportation Safety Board, which conducted the investigation, pointed to the following factors: the controller's unclear instruction to flight 954 during conditions of low visibility, in which the important word 'right' was omitted; the controller's failure to use all of the information that could have been used to confirm the location of flight 954; and the failure of the flight 954 crew to press for a clarification of the controller's instruction.[57]

In 1985, after the Tenerife accident, the Japan Research Institute of Air Law conducted a questionnaire survey of the pilots of foreign airline companies that fly into the New Tokyo International Airport in Narita. One comment from this survey pointed out that poor English pronunciation by air traffic controllers, as well as inappropriate use of terminology, also holds the potential for danger that may develop into an accident.

That comment, made by a pilot of a European airlines in his fifties who had logged from 10 000 to 20 000 flying hours, indicated that 'At times, I cannot understand the air traffic controller's words. In particular, I cannot understand when the controller uses terms other than the standard terminology. The problem is pronunciation. This can easily lead to an accident. The controllers should be trained for better English pronunciation'. There were many other comments along the same lines relating to the English of the controllers, or requesting that the controllers speak more slowly.

Concerning the Tenerife accident, the Dutch government pointed out that, as sounds similar to a football match broadcast were recorded in the background of communications with the control tower by the voice recorders of both the KLM plane and the PAA plane, the controllers may not have been concentrating their attention on what was happening on the runway.[58] In fact, looking at the results of analysis of the voice recorders of both planes, the controller even confused the call signs of the two aircraft. Although the call sign of the KLM plane was KLM 4805, the controller referred to 'KLM 80' and 'KLM 8705', and 'Papa Alpha 1736', the call sign of the PAA plane, was mistakenly given as 'Papa Alpha 7136'.

Turning attention towards matters unrelated to the job during work time often leads one to make errors or forget things. One example of this is the accident involving an Eastern Airlines flight 11 DC-9 on 11 September 1974. As the aircraft was approaching the Douglas City Airport in Charlotte, North Carolina, the crew omitted a confirmation call, in spite of following the checklist, and had a wrong conviction that the altitude of the plane was still high on the approach course, although it was actually low. Unable to see the ground surface because of ground patch fog, the aircraft crashed into a field on the approach course and burst into flames, killing 72 of the 82 passengers and crew. Analysis of the cockpit voice recorder showed that, in the midst of going through the checklist, the crew was engaged in a conversation about the Watergate political scandal.[59]

This demonstrates that a number of the same danger factors that constituted the Tenerife accident had also been experienced prior to the occurrence of another accident. However, there have been many other incidents, not widely known, which trace the same path as the Tenerife accident. In other words, accidents have occurred as the result of the same danger factors that constituted the accident at Tenerife, even after the fact.

1.13 Differences in understanding between pilot and copilot

Eight years after the Tenerife accident, on 28 May 1985, there was an accident at the Naha Airport involving an Air Self-Defence Forces MU-2 aircraft that was participating in a rescue training drill. The MU-2, beginning to take off without clearance, entered the runway from another taxiway that intersected it at a 30° angle and grazed a landing All-Nippon Airways flight 81 Boeing 747, which was moving along on the runway after having touched down.[60]

This accident fortunately resulted only in relatively light damage to the two planes and there were no injuries. It closely resembles the Tenerife accident, however, in that the Air Self-Defence Forces plane was taking off without clearance from the tower, even though the pilot was convinced that clearance had been given, in that the relationship among the flight crew did not operate to good effect, and in that the pilot had a strong latent expectation to receive clearance for take-off quickly because of rapidly changing weather conditions.

The copilot was responsible for plane control and communication. The captain was highly competent in rescue operations, while the copilot's competency was on the intermediate level. Although the captain had more total flight hours than the copilot, the two members of this flight crew were of the same rank and age, and possessed about the same qualifications, abilities, and experience. From the relationship of trust that existed between the two, an interdependent frame of mind can be inferred.

The copilot requested clearance for take-off, and, although permission for special visual flight rules was received from the controller and read back by the copilot, there was no clearance for take-off issued by the controller. Normally, the pre-take-off checklist is run through after clearance for take-off has been issued, but the captain, thinking that clearance would be given shortly, started going through the pre-take-off checklist in preparation for the take-off.

As the captain had begun the checklist right after the copilot had read back the flight clearance, the copilot had the wrong conviction that he had failed to hear the clearance for take-off, and, following the pilot, immediately began moving to the runway when the last item in the checklist was done. The captain, on the other hand, had the mistaken conviction that the copilot was beginning to move the plane out because the copilot had received clearance for take-off.

There was a difference in understanding between the captain and copilot regarding the time that the pre-take-off checklist was performed, in terms of whether it was done before or after the clearance for take-off.

1.14 Matters not clarified by the accident investigation

Accidents occur as the result of a chain of various danger factors, but is there always a relationship of inevitability, rationality, and logic between the danger factors in the chain?

Considering an accident after it has occurred, one may think that the accident was indeed inevitable, but careful consideration may reveal otherwise. Taking the factors of the Tenerife accident as an example, there was no relationship of inevitability between the KLM pilot beginning the take-off operation believing that take-off clearance had been given and the fact that, later, the KLM plane was not able to receive the transmission from the PAA plane because of noise due to overlapping transmissions. We can see that the accident resulted from the chance linking of those two factors.

If the chain of various factors were always rational, inevitable, and logical, then it would, at the very least, be possible to investigate and specify the causes of the accident by means of rational deduction based on an investigation of mechanical and material system problems. However, even if we restrict consideration to the hardware aspect, there are limits to clarification of the process by which an accident occurs through rational and logical inference based on the results of an accident if the accident or incident involves unsafe events that arise through a chain of events that cannot be arrived at using ordinary logic alone.

As a specific demonstration that this is true, let us look at an actual example of an extremely dangerous situation concerning aircraft maintenance taken from a report to the Japan Research Institute of Air Law. When a DC-10 aircraft was landing at Osaka Airport, the thrust reverser of the No. 2 engine became inoperative. Therefore, in accordance with the Aircraft Maintenance Manual, procedures were implemented before the plane flew on to the New Tokyo International Airport in Narita to lock the engine translating sleeve in the forward thrust direction so that the inoperative thrust reverser could not operate inadvertently. (The translating sleeve is an annular cover on the rear part of the engine that closes the reverse engine opening when the thrust is in the forward direction – see Fig. 8).

Because those procedures satisfied the operating limitation standards,[61] the DC-10 took off from Osaka to head towards Narita. However, the climb was slower than usual, and, when the engine output was increased and the climb continued, the No. 2 engine reverse thrust warning light soon lit up. Understanding that the No. 2 engine was in reverse thrust, the captain shut down the No. 2 engine[62] and turned back to Osaka Airport.

The procedure implemented before the flight to Narita was mechanically to lock the thrust reverser after removing the electrical connector that sends the electrical signal to the thrust reverser, without disconnecting the motor. According to the Aircraft Maintenance Manual of the DC-10, the thrust reverser must be locked by locking four of six places on one engine.[63] However, only two places had been secured, and it was discovered that those two places were broken and that the translating sleeve had moved to the reverse thrust position.

Also, a detailed inspection of the removed thrust reverser actuating motor revealed that the selector valve, which switches the direction of flow of the pneumatic air that turns the motor axle, was open in the direction of reverse thrust. There is no particular procedure for whether the flow should be in the forward thrust direction or in the reverse thrust direction when the electric plug is pulled out or for confirming that situation. That is because there is a shut-off valve upstream of the selector valve and, when that shut-off valve is completely closed, the pneumatic air cannot usually reach the selector valve.

By misfortune, however, a 0.5 mm snippet of safety wire (a wire that prevents a nut and bolt from loosening as a result of vibration during flight) had jammed in the shut-off valve, creating a gap through which pneumatic air could leak through to the selector valve. In the task of disconnecting the duct, placing a dust cover on the open end of the duct

Translating sleeve in the reverse thrust position
(modification of a diagram from "Prevention of Accidents in Large-scale Systems II")

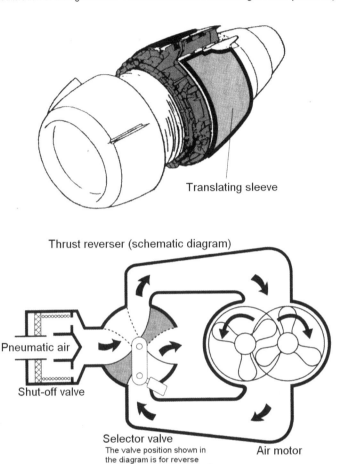

Fig. 8 Engine translating sleeve and thrust reverser

is a basic maintenance procedure. In the process of locking the thrust reverser, however, the duct is not cut off, so the piece of safety wire could not have entered the duct as a result of that process. That is to say, what can be imagined is that the piece of safety wire got into the duct when someone made the minor error of not applying the dust cover in a previous duct disconnection, and then moved to jam in the shut-off valve. That in itself is a rare occurrence.

Pneumatic air supplied by the engine as the plane made its climb gradually escaped through the slight gap created by the jammed piece of safety wire and, because the downstream selector valve was open in the reverse thrust direction, the pneumatic air flowed in that direction. This pneumatic air caused the lockplate that was preventing the operation of the thrust reverser in two places to break, thus causing the thrust reverser of the No. 2 engine to operate.

Three factors contributed to this incident: the fact that there was a gap in the shut-off valve from a jammed piece of safety wire, the fact that the electrical connector was pulled out when the selector valve for the air motor was in the reverse thrust position, and the fact that the thrust reverser was locked in only two places rather than in four places as it should have been.

If the piece of safety wire had not entered the duct, or if the selector valve had been opened in the forward thrust position, the pneumatic air would not have flowed in the reverse thrust direction. Also, if the lockplate had been attached in four places rather than in only two places, the thrust reverser probably would not have functioned even if the pneumatic air did flow in the reverse thrust direction.

This example shows how an earlier minor error in maintenance, unrelated to the procedures implemented to disable the thrust reverser, linked with a later factor to cause an incident. This, therefore, also realistically demonstrates the possibility that a minor error that cannot rightly even be called negligence can link together with other later factors to cause a major accident.

Luckily, because the piece of safety wire that jammed was very thin and small, the movement of the thrust reverser was gentle and left sufficient margin for the pilot to shut down the engine and turn back to the airport from which the plane had taken off, so the incident did not reach the stage of becoming an accident. Because of that, none of the other parts were damaged and the engine could be examined in a static state, so, when the broken driveshaft was replaced, it was possible to discover that the piece of wire had jammed in the shut-off valve.

If, however, the piece of safety wire had been a little larger, and thus the gap in the shut-off valve had been a little wider, causing the thrust reverser to begin to function abruptly, there would have been a high danger of the plane becoming uncontrollable. Particularly during the take-off run, when the engine output is at maximum, and so the reverse thrust would also be maximum should the thrust reverser start working, there would be an extremely high possibility of the aircraft crashing immediately after lift-off.

If a crash had occurred, damage to the aircraft would probably have made it quite impossible to discover how the fragment of safety wire had contributed to the cause of the accident. In that case, it might have been inferred that the accident had occurred because the lockplate had broken owing to deterioration over time, that the stow latches (hardware for holding stored objects in place) came loose owing to vibration during flight, or that the thrust reverser had moved backwards owing to external air pressure during flight.

This also shows how we can fail to see clearly all of the causes of an accident by simply accepting a single factor or the immediate factor after the accident has occurred. There are many cases, then, in which the chain of various factors is not inevitable, rational, and logical, but rather accidental, irrational, and illogical.

1.15 Accident investigations conducted by reverse logic

Whenever a major accident occurs, regardless of the field, there is a cry for the prevention of a recurrence. The approach taken for that purpose has generally been to investigate the causes of major accidents and to derive preventive measures based on the results of those investigations.

Accident investigations involve a rational analysis of accidents that are the result of various danger factors that are linked and conflict in various forms. Accordingly, if the cause of an accident is limited to a mechanical system defect that arises from a completely rational and logical chain of events, then it is probably, to some extent, possible to work out that cause by rational deduction, for example based on a scientific analysis of plane wreckage. However, when the parts of the aircraft are crushed or damaged by fire or explosion, there are limits to the ability to determine causes related to mechanical systems, as was explained for the maintenance-related problem in the DC-10 engine described in the previous case.

Nevertheless, if the chain of danger factors is not logical, even though it may in fact be nearly so, unless there is hard evidence, that chain cannot be specified as the cause of the accident. In such a case, the only available method is to make a deduction based on the logical probability (likeliness or believability) of the overall conflict and chaining of danger factors. This is because investigation reports that contain what appear at first glance to be reckless inferences generate doubts concerning objectivity, and so have a low probability of social acceptance. As a result, the persons responsible for the report are afraid that incorporating such 'extraordinary inferences' would reduce the credibility of the report. Expressed in another way, in determining the cause of an accident by means of an accident investigation, it is generally not possible to step outside the bounds of inferences that 'seem to be the most probable' from looking at the results of an analysis of the circumstances at the accident site and the wreckage.

Looking more closely at this limitation of accident investigations, let us consider the example of an accident in which a Japan Airlines plane crashed into the sea near Haneda on 9 February 1982. This accident again demonstrates that a pilot who is sound in mind and body at the controls of the aircraft is a basic prerequisite to ensuring aviation safety. While this was a terrible accident in which 24 of the 174 passengers and crew members died, what was alarming to the public was the astounding cause of the accident. Although the term 'reverse thrust' had become popular in the mass media, the direct cause of the accident was the truly unbelievable act of the pilot pulling the reverse thrust lever just before landing. The Aircraft Accident Investigation Commission of the Ministry of Transport Japan, which investigated this accident, stated: 'It is concluded that the immediate cause of this accident was an abnormal operation by the pilot as the result of a psychological abnormality'.[64]

Considering the limitations of accident investigations that have been described already, one might feel some doubt about whether the conclusion of this accident investigation report, which might normally be considered a little extraordinary, would have been possible had the entire flight crew survived. People who were associated with the pilot had been aware of his psychological problem for a number of years prior to the accident, but would the kind of inference made in the accident report be possible if the entire crew had died? Would it not have been difficult for other pilots, to say nothing of the general public, to accept a psychological problem that would end a person's own life?

Let us assume that the pilot had never in the past been seen to have psychological problems, but had pulled the reverse thrust lever because of a sudden psychological problem, and that the crew who had witnessed the act had all died. What would the Accident Investigation Commission have determined to be the cause of the accident? Of course, it would probably have been possible to determine from an examination of the plane wreckage that the cause of the accident was the reverse thrust lever being pulled for some reason, but the matter of why it had been pulled probably would have remained a mystery.

1.16 Forward-looking information obtained by IRAS

In the case of major aircraft accidents, the pilot and others who are in a position best to know the process that leads up to the accident often lose their lives.   Even when the pilot survives (and this is true of the maintenance personnel, air traffic controllers, and other such persons concerned with the aircraft involved in the accident as well), it is difficult to expect honest testimony from such persons, given that these individuals may be facing criminal charges, such as professional negligence resulting in bodily injury or death[65] or danger to an aircraft.[66] Even if criminal charges are not laid, there may be some form of administrative punishment such as revocation or suspension of licenses,[67] or punishment by employers such as demotion or removal from the flight schedule as a penalty, all of which are accompanied with reduction in income.

Weather conditions greatly affect the operation of aircraft, but, with the exception of highly unusual cases of weather phenomena that have never before been experienced, human error is involved in the occurrence of dangerous events at one stage or another.   It would be reasonable to say, however, that it has never really been possible to achieve an accurate and detailed investigation of human factors, and it never will be possible.

It is true that the accident report conclusions arrived at by logical deduction are useful in making improvements in hardware.   Such results, however, are useful only for accidents that arise from a logical rational process; they are entirely useless for irrational non-logical chains of danger factors that deviate from the deductive process of the investigation.   Furthermore, inasmuch as the accident investigation does not begin until after the accident has occurred, it has definite limitations from the viewpoint of safety securing.   Moreover, the extent to which the interests of the individuals and the governments that are concerned with the accident are at play increases with the size of the accident, and it is more likely that 'other considerations' may compete with the primary considerations.   As illustrated by the Tenerife accident, we could say that the accident causes presented as the conclusion of an accident report are no more than the opinion of the accident investigation board that investigated the accident.   Moreover, even granting for the moment that the investigation discovered the true causes of the accident, it is still not possible to identify the position of those danger factors, including the 'soft' factors, among the many potential danger factors of that field.

The identification of the true causes of an accident by means of an accident investigation is, from the beginning, an extremely difficult matter.   Particularly in the prevention of accidents in large-scale complex systems, in which human factors play principal roles, although the conventional method of 'learning from past accidents (the so-called 'tombstone safety' approach)'[68] is also important, a more positive method that gets to the root of an accident must be devised.

To this end, the author has proposed the Incident Reports Analysing System (IRAS). The goal of this system is to prevent accidents by collecting large amounts of information on the incidents that occur prior to accidents, and analysing those data to search out the latent danger factors from a macroscopic perspective. The system is described briefly and its import is discussed in the following sections.

1.17 Disasters preventable through conveyance of information

The radiation leak at the Three Mile Island nuclear power station near Middletown, Pennsylvania, in 1979 is often cited as an example of an accident that could have been prevented if information on exactly the same kinds of incident that had occurred prior to the accident had been made available.

While that accident is said to have begun with a mechanical failure that was aggravated by human error to develop into a major accident, it is known from hearings held during the investigation after the accident that many minor accidents and problems that did not reach the stage of becoming accidents had occurred prior to the radiation leak. Furthermore, 11 of those minor accidents involved illegal actions, and the fact that they were not dealt with appropriately had the effect of inviting a major accident.[69]

Also, in the field of aviation, the same kind of example can be seen in the crash of the American Airlines flight 191 DC-10-10 on 25 May 1979. All 271 of the passengers and crew died in the crash and two persons on the ground were killed by flying debris, in what can be called the worst air accident in US history. The US Federal Aviation Administration suspended the aircraft type certificate for the type of aircraft involved, grounding the planes for 37 days and exerting a great influence on worldwide aviation.

This accident occurred when the engine and pylon (the structural member that attaches the engine to the wing) on the left side of the plane fell off, damaging the oil pressure system and causing the plane to crash into the ground, resulting in a massive explosion. The investigation of the accident concluded that the reason that the engine fell off was cracking that had developed where the engine was attached. Those cracks expanded under operating load until the residual strength decreased to the critical value and the engine fell off.[70]

The manufacturer of the aircraft, the McDonnell Douglas Corporation, had advised all airline companies that the procedure for removing the engine from the aircraft for maintenance is to remove the engine and pylon separately. However, two airlines, American Airlines and Continental Airlines, considered the procedure of removing the engine and pylon from the plane together as a unit to be more efficient, and implemented that procedure. While that procedure can certainly shorten the maintenance time, the work is delicate and there is much risk of creating sources of metal fatigue at the time of removal and reattachment.

Prior to this disastrous accident, in December 1978 and in February 1979, cracks had been discovered in the engine attachment parts of two DC-10 aircraft of Continental Airlines, a company that employed the same engine removal procedure as had American Airlines. That cracking resulted from rough handling when the engines and pylons were removed together. In the case of Continental Airlines, the cracks were discovered during aircraft maintenance, and a major accident was averted because appropriate repairs were made. That information, however, was not conveyed to American Airlines and so the disaster of 25 May 1979 was not averted.

Because this was a maintenance-related problem, Continental Airlines had no obligation to report it to the Federal Aviation Administration, but they issued an Operation Occurrence Report (OOR) and informed other airlines of the problem. However, there was no description in that report concerning the cause of the damage. Receiving the report, American Airlines filed it into a distribution list, but that information did not reach the persons in charge of pylon maintenance.

Although this fact came to light after the American Airline flight 191 disaster, if the information had been conveyed accurately and in detail to American Airlines, which employed the same maintenance procedure, it would have at the very least demanded greater care to be taken during the engine removal process, and so there was a possibility that the accident could have been prevented. This illustrates the need to construct a system that will make it possible to report factual information concerning incidents and accidents at a level above the corporation or the state.

In our investigations, too, it has become clear that the same types of unsafe event have occurred prior to the occurrence of accidents. Several years before the accident in which a Japan Asia Airways flight EG 292

DC-8 hit approach lights while landing at Naha Airport on 19 April 1984,[71] there was an accident in which, on the same runway of the same airport, control of the plane was lost momentarily because of a wind shear (a sudden violent change in wind direction and speed) and the tyres of the plane burst as it touched down with excessive force. Exactly the same situation had been reported previously in the 10 January 1988 accident in which the Toa-Kokunai Airlines YS-11 aircraft overran the runway and plunged into Lake Nakanoumi during take-off in light snow from Yonago Airport.[72]

The principle behind Heinrich's law is also at work in the field of aviation, and major accidents do not generally just suddenly occur one day with no forewarning at all. There is no question that serious accidents are preceded by many occurrences of similar minor accidents and lesser problems. The US National Transportation Safety Board, which investigated the American Airlines flight 191 accident, stated in their report that 'It is noted that the Federal Aviation Administration and American Airlines did not know that damage to the engine attachment parts of the two Continental Airlines DC-10s had occurred until after the accident involving flight 191', and went on to point out that 'The current reporting system has limitations'.

If there were a system for collecting such incident information by a special organization and accurately and rapidly communicating that information to all of the companies and organizations within a given field, then that incident information could be used in the prevention of accidents throughout the entire field. When such incidents are correctly understood and reported, without oversights, and appropriate measures are devised, then it will become possible to prevent accidents even in large-scale complex systems.

1.18 Advances in science and technology and new kinds of danger

That advances in science and technology bring about new kinds of danger has already been pointed out in the discussion of the Chernobyl nuclear accident, but what about the field of aviation?

At the same time that the second-generation jets appeared on the scene, the Inertial Navigation System (INS) was developed. Connecting that device with the automatic piloting equipment enabled not only a reduction in the pilot's workload (the number of tasks performed per unit time) but also very highly accurate automatic navigation. We are now seeing new types of danger, however, that were not present in earlier navigation methods, such as celestial navigation, in which position is calculated from measured angles to stars relative to the horizontal, and Loran, in which positions are determined from the difference between the times that radio signals from two or more radio stations take to reach the aircraft.

A typical example is the incident that occurred on 30 November 1985. A Japan Airlines flight 441 Boeing 747 encountered cumulonimbus offshore near Niigata while flying over the Sea of Japan on a flight from Narita International Airport to Moscow, and the pilot switched to the heading mode (an autopilot mode in which the set heading of the aircraft is maintained) to avoid the cumulonimbus.   Afterwards, however, the pilot forgot to return the mode selection switch to the INS mode, and so the aircraft proceeded north, well out of the air way, causing a number of Soviet military aircraft to be scrambled.[73]

The INS requires that the current position be input and the plane remain still with the mode switch in the 'align' position (gyroscope stabilization mode) for a certain amount of time to stabilize the gyroscope and increase accuracy, and that the switch be set to the 'navigation mode' position before the plane begins to move. That is because, if the plane is moved while the INS is set to the 'align' position, the stability of the gyroscope cannot be guaranteed and the accuracy of navigation is reduced to a remarkable degree.

This introduces the possibility for new types of dangerous error that could not have arisen in the older methods of navigation, such as beginning to move the plane before consciously setting the switch to the navigation mode or while having forgotten to set the mode, or entering the current position incorrectly. The scope of automatic control is being increased on the basis of the assumption that humans are inclined to make errors, but this brings with it new types of danger.

The accident involving the landing China Airlines flight 140 Airbus 300-600 (high-tech model) at Nagoya Airport on 26 April 1994 demonstrated the danger of a stall and crash resulting from the discord that arises between man and machine owing to conflict between the design approach that aims for automation and computer control through the development of new technology and the reflexive operation by a human pilot.

Since the time powered flight was first realized, the most basic and reflexive operations for a pilot during final approach are to push forward on the control column when the nose of the aircraft rises and to pull back on the control column when the nose goes down.

Usually (always, in the case of US aircraft) there is a mechanism that automatically disengages the automatic control system when the force applied to the control column exceeds a certain limit. In the Airbus 300-600, however, the automatic control system is not disengaged even if external pressure is applied to the control column, to point the nose down, for example; it continues to operate, counteracting the external force by operating the horizontal stabilizer so as to raise the nose.

According to the report of the Aircraft Accident Investigation Committee of the Ministry of Transport of Japan, the copilot of the plane inadvertently activated the 'go lever' (landing abort lever) during a manual control approach to the airport, placing the aircraft in the go-around mode. After that, the automatic control system was engaged, so that, when the pilot pushed forward on the control column to lower the nose and continue the approach, the automatic control system (i.e. the computer) reacted by setting the horizontal stabilizer to the maximum position for raising the nose of the aircraft. As a result, the automatic stall prevention system was activated, increasing the engine output and further raising the nose. Instead of recovering from the stall, the aircraft plummeted downwards, crashed into the ground, and burst into flames. This accident resulted in the deaths of 264 passengers and crew members and serious injuries to seven passengers.[74]

Before this accident occurred, at least three serious incidents of a similar type involving the Airbus 300, including the sister series Airbus 300-600, had been reported to the Airbus Corporation. Those incidents had occurred on 1 March 1985, 9 January 1989 (Finland Airlines, during approach to Helsinki Airport), and 11 February 1991 (during approach to Moscow Airport).[75] A similar unsafe event also occurred just after the China Airlines crash on 24 September 1994, during approach to Orly Airport in Paris.

Although these incidents had various causes, they are all similar in that the pilot's operation of the control column and the action of the automatic control system were at odds, and the nose up-down balance of the aircraft was effectively lost (a situation referred to as 'out of trim'). The pilot, unable correctly to understand the situation, felt it necessary to react to the sudden change in the attitude of the aircraft.

Each of these four incidents occurred at high altitude, so it was possible to recover from the situation and, fortunately, the stage of an accident was not reached. The Airbus Corporation, on each occasion, published the fact as an attention item in the Flight Crew Operating Manual (FCOM). However, there was insufficient technical explanation, examples of what may happen or how to recognize problem situations, and systematic explanation of the operating procedures. Furthermore, there was no training in responses to such problems. The captain of the plane involved in the accident, from past flight experience with the Boeing 747, may have believed that manual operation would have priority over automatic control when the control was operated by hand.

Also, the Airbus Corporation had recommended an improvement of the automatic control system in the July 1993 'Service Bulletin', technical information that is issued to all airline companies. The recommended improvement was that the automatic control system should be automatically disengaged when the system is in the go-around mode and a pressure greater than 33 lbf is applied to the control column at altitudes of over 400 ft above ground level. However, because that improvement was not mandatory, but only a recommendation, China Airlines set plans for making the revision but did not judge it to be an urgent matter. Because of that, the reported incident information was not fully applied, and the accident occurred. If the improvement had been made, the accident would not have happened.

This does, however, show that there are incidents associated even with these new types of danger that accompany scientific and technological progress, and, if those incidents are quickly understood and appropriate measures taken, such accidents can be prevented.

1.19 The pitfalls of computer control

The control systems of high-tech aircraft are referred to as 'fly-by-wire' systems, because the control column and control surfaces are connected by electrical wires or optical cable via a computer rather than mechanically by hydraulic systems or cables. The input from the control column is processed together with various data by the computer and converted to a signal that controls the movements of the control surfaces, so there is no mechanical link between the control column and the control surfaces. Accordingly, the pilot's intent is changed by the computer program. To the pilot, the computer program is a perfect black box that, at times, acts in opposition to the pilot's intention. The computer is constructed so that, if it loses control, automatic control is automatically disengaged, but it is designed in that way on the basis of the recognition that the computer may indeed lose control. When the automatic control system is acting against the pilot's intentions and cannot be released, the signal from the computer and the aircraft's response to the signal may become out of phase, creating an oscillation, which may gradually increase in amplitude. This is a problem that dogs automatic control systems, and is called Pilot Involved Oscillation (PIO). It has been the cause of accidents in which YF-12 military aircraft have crashed.[76]

Also, on 8 June 1997 the Japan Airlines flight 706 MD-11 plane experienced a severe up and down movement during descent over Mie Prefecture in Japan. Although there was no damage to the aircraft at all, 12 persons suffered light or serious injuries (those with serious injuries included one passenger and three crew members working in the passenger cabin, one of whom died 20 months later). While the plane was proceeding from Kushimoto towards Nagoya under automatic control at 350 knots (about 390 mile/h) and descending, the speed suddenly increased by 20 knots (about 22 mile/h) in 20 s, exceeding the maximum operating speed V_{mo}. The pilot tried to operate the automatic control so as to reduce the descent rate and reduce the speed, but the pilot's intention differed from the computer program, and, even though the speed limit was being exceeded, the automatic control responded very slowly so as not to discomfort the passengers. Therefore, the pilot was compelled to apply the speed brake. Immediately after that, the automatic control disengaged and the nose of the aircraft began to oscillate up and down.

There is no specified procedure for dealing with this kind of situation where speed limits are exceeded. All the pilot can do is to try various ways to hold down the speed. Thus, there are still many more problems that must be solved for high-tech aircraft.[77]

The MD-11 is in about the same generation of development as the Boeing 747-400 series. It takes automation one step further, and has characteristics that are remarkable compared with other aircraft. One such characteristic is that, during flight, fuel carried in the forward part of the wings can be moved to tanks in the horizontal tailplanes, thus moving the centre of gravity of the aircraft towards the rear to reduce fuel consumption while cruising. When the centre of gravity is closer to the rear of the plane, the control column operation becomes lighter, but the plane becomes less stable. To ensure normal stability when landing, the fuel in the tail tanks is returned to the forward wing tanks. In the case of a sudden descent, however, it has been pointed out that the transfer of fuel to the forward tanks may be too slow, and there is thus a possibility of an extremely unstable condition.[78]

Other aircraft already employ a yaw damper (an automatic control device for direction only) so that the vertical tail fin can be made smaller, thus increasing economy of operation by reducing air resistance during flight. The MD-11, assuming computer control, also reduces the size of the horizontal tailplanes. The horizontal tailplanes on the MD-11 are about 70 per cent the size of the tailplanes on the DC-10, for example.

On 5 September 1997, the Aircraft Accident Investigation Commission of Japan issued the following progress report regarding the MD-11 accident: 'Immediately after the automatic control was disengaged, there were five sudden oscillations in the pitch of the aircraft, the cause of which may have been "PIO", a phenomenon in which the aircraft undergoes an oscillation counter to the intentions of the pilot'.[79]

A similar incident was experienced by American Airlines flight 107 MD-11 on 29 June 1994. While the plane was cruising under automatic control, the auxiliary copilot was sitting in the pilot's seat because the pilot was on break and the copilot was sitting in the seat on the right with his legs crossed. When a cabin crew member brought in a drink for the pilot in a box, the auxiliary copilot said that he was setting it on the base of the observer's seat, but could not easily place it there. He thought that if the copilot's seat were moved a little further forward, he could set the box in place, so, without saying anything to the copilot, he pressed the switch to move the seat forward.

As a result, the copilot's crossed legs hit the control column, disengaging the automatic control system. In response to the forward pressure on the control column, the nose of the aircraft suddenly dropped and then continued to move up and down. Although there was no resulting damage to the aircraft, 97 passengers and crew were subjected to a negative G force and 17 sustained serious and minor injuries.[80] The copilot brought the plane back to level flight by manual control, but new types of hazard that have never been seen before arise when the automatic control system is automatically disengaged because the operation of the aircraft goes beyond the range preset for automatic control.

Although about 200 MD-11 series aircraft have been commissioned, more that 18 accidents in which there was up and down oscillation or in which the tail of the aircraft hit the runway during landing because the nose of the aircraft rose suddenly in opposition to the intended operation are said to have happened by the end of 1997.

The FAA and the McDonnell Douglas Corporation (currently The Boeing Co.), the manufacturer of the MD-11, are reported to be studying ways to increase the up and down stability in the nose of the MD-11. Examples of these methods are improving the computer software, changing the pitch rate damper, changing the response of the aircraft to the control column force, and adding an elevator lock-out (a restraint that stops operation).

1.20 Where do the relevant danger factors lie?

In an examination of analytical methods for unsafe events that involve human factors, we find that there has been no attempt to collect a large number of incident reports, and that there has been no suitable method of analysis available. In the following, I will outline a newly developed analytical method called the Multidimensional Analysis of Incident Reports (MAIR) (see Fig. 9), which is premised on the IRAS that we have established.[81] This method may be somewhat difficult to grasp, but I will describe it simply here to make it easier to understand the specific examples and the overall results.

Firstly, any task (job) is considered to be a series of information processing activities that are divided into four phases: information source, information receiving, decision, and action or instruction. The results of confirming the action (operation) or instruction phase and

---

MAIR
(Multi-dimensional Analysis of Incident Reports)

Information Processing Activities
      Phase I   Information source
      Phase II   Information receiving
      Phase III   Decision
      Phase IV   Action / instruction

Error Modes

Background Factors
      Personal factors
      Inter-personal relationship factors
      General factors
      Special background factors (special circumstances)

Reason for Recovery

Degree of Danger (degree of approach to accidents)
      Virtually none
      Fairly little
      Considerable
      Serious

Incident Critical Event and Phase

Fig. 9 Outline of MAIR

new information become the next information source, and this cycle repeats (see Fig. 10).

For example, let us consider the pilot's task of averting a near miss according to this approach. The information flow begins with looking to the outside of the plane to see the other aircraft (information source), followed by the discovery of the other plane (information receiving), the decision to avoid (decision), and then operation of the control column (action or instruction). Any kind of job can be considered to consist of repetitions of these four phases, and incidents and accidents can be considered to consist of one or more events that represent flaws in one of these four phases.

Secondly, it can be seen that there are various types of error that a person can commit in job performance. Technically, these are referred to as 'error modes', and can be classified into nine categories. The classification is based on how 'what was actually done' deviated from 'what was required'. That is to say, the way in which what was actually done deviated from what was required to be done can be classified into nine categories, including the appropriate behaviour (i.e. no deviation). We can try to express the error mode for each phase in the following way:

1. Appropriate response (this is not an error – it is included here because it is necessary for the analysis).
2. Lapse of memory; forgetting to do something that is required. This is manifested as 'did not see' or 'did not hear (no input)' in the information receiving phase, as 'forgot (no decision)' in the decision phase, and as 'forgot (no action/instruction)' in the action or instruction phase.
3. Low level of awareness. This is manifested as 'failure to see', 'failure to hear', or 'failure to notice' in the information receiving phase, as 'easy-goingness/assumption' in the decision phase, and as 'sat idle/ aimlessness', 'meaningless action', or 'instruction not received clearly/unclear pronunciation' in the action or instruction phase.
4. Mistaking what must be done. This is manifested as 'saw incorrectly', 'heard incorrectly', or 'mistaken impression' in the information receiving phase, as 'wrong/mistaken conviction', 'hasty conclusion', 'one's own interpretation' or 'wrong expectation/ mistaken expectation' in the decision phase, and as 'incorrect action/instruction', 'mistaken or inappropriate action/instruction' or 'incorrect speech/misspeaking' in the action or instruction phase.
5. Uncertain about what must be done. This is manifested as 'vacillation' in the information receiving phase, as 'vacillation' in

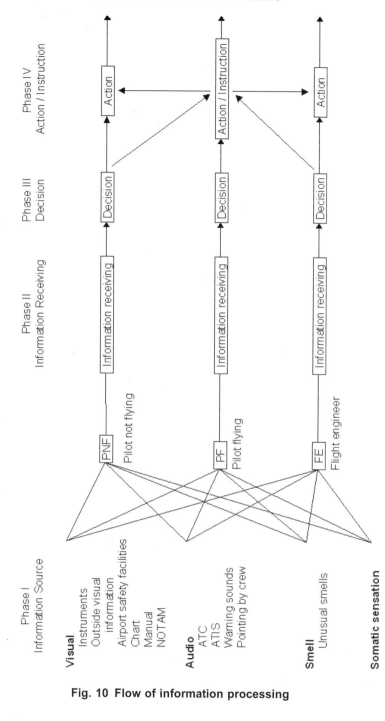

Fig. 10 Flow of information processing

the decision phase, and as 'loss of balance/stability', 'lack of effective teamwork', 'confusion', or 'ambiguous instruction/vague instruction' in the action or instruction phase.

6. Focusing on only one of multiple things that must be done at the same time. This is manifested as 'excessive focus/excessive concentration on one point' in the information receiving phase, as 'fixation' in the decision phase, and as 'fixation' or 'persistent action/instruction' in the action or instruction phase.

7. The actual behaviour is excessive. This is manifested as 'preconception' in the information receiving phase, as 'unreasonableness', 'high-handedness', or 'excessive thoughtfulness/thinking too much' in the decision phase, and as 'recklessness', 'forcefulness', 'excessive action/operation/work', or 'overdoing/overtightening' in the action or instruction phase.

8. The actual behaviour was not carried out by the appropriate time or to a sufficient degree. This is manifested as 'delay' in the information receiving phase, as 'delay' or 'insufficient consideration' in the decision phase, and as 'delay', 'incomplete action/instruction', or 'insufficient action/instruction' in the action or instruction phase.

9. Proper action impossible owing to malfunction or obstacle. This is manifested as 'could not see clearly', 'could not see', 'could not hear clearly', 'could not hear', 'could not reach', 'hard to reach', or 'could not reach sufficiently' in the information receiving phase, as 'difficult decision' or 'inability to decide' in the decision phase, and as 'difficult action or instruction' or 'inability to act or work' in the action or instruction phase (see Table 1).

Then, by placing these nine categories along the vertical axis of a chart and placing the three information processing phases (excluding the information source phase) used in the above description along the horizontal axis, and then plotting the error modes of individual incidents on that chart, it is possible to draw an overall image of the flow of the error modes. That is to say, the way that a task should be performed in each information processing phase is analysed event by event in terms of which of the nine categories applies.

For example, in an incident in which a near miss occurs because the pilot for some reason cannot avoid the other aircraft, the following three cases can be considered (of course, there are various other cases as well):

1. The other plane is discovered appropriately (information receiving), the decision to evade is also made appropriately (decision), but the

Table 1 Modes in each phase of information processing activities

Category	Actual direction of human energy	Phase II Information receiving	Phase III Decision	Phase IV Action/instruction
1		Appropriate	Appropriate	Appropriate
2		• Did not see • Did not hear (no input)	Forgot (no decision)	Forgot (no action / instruction)
3		• Failure to see • Failure to hear • Failure to notice	Easy goingness / Assumption	• Instruction not received clearly / unclear pronunciation • Sat idle / Aimlessness • Meaningless action
4		• Saw incorrectly • Heard incorrectly • Mistaken impression	• Wrong conviction • Hasty conclusion • One's own interpretation • Wrong expectation / mistake in expectation	• Incorrect action / instruction • Mistaken or inappropriate action / instruction • Incorrect speech / misspeaking
5		Vacillation	Vacillation	• Loss of balance or stability • Lack of effective teamwork • Confusion • Ambiguous instruction / vague instruction
6		• Excessive focus • Excessive concentration on one point	Fixation	• Fixation • Persistent action / instruction
7		Preconception	• Unreasonableness • High-handedness • Excessive thoughtfulness / thinking too much	• Recklessness • Forcefulness • Excessive action/operation/work • Overdoing / overtightening
8		Delay	• Delay • Insufficient consideration	• Delay • Incomplete action / instruction • Insufficient action / instruction
9		• Could not see clearly / could not see • Could not hear clearly / could not hear • Could not reach sufficiently / hard to reach / could not reach	Difficult decision / inability to decide	• Difficult action or instruction • Inability to act / work

Explanation of symbols:
(→ indicates the direction of human psychological or physical energy (what was actually done);
● represents the object to which human energy must essentially and absolutely be directed so as to prevent an incident or accident (the object of what was required);
○ represents a non-essential object.
After Miyagi, 1992, Proc. BICMRS, International Academic Publishers.

avoidance manoeuvre (action or instruction) is late, so the near miss occurred.

2. The other plane is discovered appropriately (information receiving), but the pilot vacillated in making the decision to evade, and, as a result, the manoeuvre was delayed and the near miss occurred.

3. The other plane is difficult to see, and, as a result, the decision to evade was delayed, and, in turn, the manoeuvre was also delayed and so the near miss occurred.

This is to say that, when examined in detail, a near miss may involve various cases. To begin with, performing this kind of analysis makes it possible correctly to understand by means of a time series what kind of information processing flow is involved in the near miss and at which phase the error is made and what kind of error is made.

1.21 The phase that determines the occurrence of an incident

Excluding psychological disorder or intentional acts, on-the-job errors made by professionals whose skills are above a certain level generally have corresponding background factors. That tendency becomes even stronger when such persons receive the same kind of training. Without measures for improvement with respect to the background factors, no increase in safety can be expected, no matter how severely the person concerned is dealt with. There is a high possibility that other persons will make the same error when they encounter the same situation.

Incident information is not a question of who created an unsafe event, but rather has the meaning of understanding the errors that people make under those circumstances. The gathering of incident information that deals with human factors is by no means intended for fixing responsibility for causing the incident on the person concerned; its purpose is the prevention of accidents. Accordingly, rather than stopping with the kind of analysis described above, the original objective of accident prevention cannot be attained unless the background factors that lead to the error made by the person concerned, that is to say, the latent danger factors, are found and improvements for eliminating or circumventing those latent danger factors are investigated.

Therefore, examining the three patterns of near misses, we first see that each pattern has a phase that determines the incident (i.e. an incident critical phase), as follows:

1. Delay in the avoidance manoeuvre in the action or instruction phase.
2. Vacillation in deciding to evade in the decision phase.
3. The other plane was difficult to see in the information receiving phase.

To simplify the explanation, each of the above examples consisted of a single event, but one incident most often comprises multiple events. In such cases, one of those events determines the occurrence of the incident. That is referred to as the incident critical phase. However, it is meaningless to point to the errors made in the incident critical phases and to say to the pilot 'Take care not to make these errors'. There are always background factors behind human errors. Then, unless those background factors change or improvements are made to eliminate or avert those factors, those same errors will be repeated and sooner or later a major accident will occur.

  Let us suppose that we search the incident reports for the respective background factors of those errors and obtain the following kinds of information:

1. While flying under automatic control, the pilot, because of fatigue, did not first disengage the automatic control before performing the evasion operation manually, thus delaying the operation.
2. At night, in particular, the other plane can be seen only as a point of light in the darkness, and it is not easy to judge its altitude, direction of flight, distance, etc., thus inducing vacillation in the pilot's decision.
3. While flying westward towards the setting sun, even though the weather conditions were good, the light from the sun made it difficult to see the other plane.

In these cases, the presence of the background factors of fatigue, the darkness of night, and back-light from the sun delayed the pilot's evasion operations, causing the near miss to occur. That is to say, these background factors can be considered to be the latent danger factors of the respective incidents.

  If it is possible to clarify the errors that determine the occurrence of incidents and the latent danger factors that lie in the background of human errors, then we could derive the general correlation between the errors and the latent danger factors from an overall consideration of the results of analysing many incidents. Here, each incident had only one associated latent danger factor, but actually cases in which there are multiple contending factors, such as long flight times and looking towards the sun, and cases in which events are linked, are more common.

That is to say, we can grasp the probabilities involved in 'the kinds of mistakes people make under certain circumstances' (the correlation differs with the field and also the time of the investigation), and only when we are able to draw out these probabilities can we study and propose measures to eliminate the danger. Quantification method III, a statistical method of multidimensional analysis, will be useful in obtaining a comprehensive understanding of the correlation between errors and latent danger factors (see Appendix A for an explanation of quantification method III).

The background factors (latent danger factors) that lead to errors are not assumed by the analyser, they must be found in the descriptions in incident reports. Innovation in the field of aviation is steadily moving forwards. It is natural that the background factors differ according to the time of the investigation, but it is also necessary to warn against prediction and inference on this basis, because the basic stance of respect only for the facts may be spoiled if there is room for the intervention of the preconceptions of the analyser.

1.22 Evaluating the degree of danger

For these incidents, the degree of danger was taken to be the probability of an accident occurring.

The degree of danger can be divided into four stages: 'serious', 'considerable', 'fairly little', and 'virtually none'. Incidents are ranked by degree of danger according to how close the incident came to becoming an accident under the specific circumstances of the incident. Differences in error mode, whether or not manual rules were violated or the degree of such violation, and the extent of loss or amount of financial losses are not issues in that ranking.

Because the purpose of IRAS is to prevent accidents to the greatest extent possible, the evaluation of the degree of danger is meaningless if it is not based on the probability of development into an accident or closeness to an accident. At the same time, however, it should not take into account the responsibility of the persons involved for a situation developing into an accident. For example, it cannot be said that incidents for which the error mode is 'unreasonableness/high-handed-ness', 'recklessness', or 'forgot' have a higher degree of danger than incidents for which the error mode is 'failure to see', 'failure to hear', 'could not see clearly', 'could not see', 'could not hear clearly', or 'could not hear'.

However, because the degree of danger is an evaluation of how close an incident comes to becoming an accident under specific conditions, in the case of forgetting to return the Inertial Navigation System (INS) mode switch to the original position after a manoeuvre to evade cumulonimbus while flying over the ocean in an airspace where there is no radar monitoring, the pilot cannot expect advice from air traffic control unless the pilot himself notices the problem.

In the composite air routes of the North Pacific Ocean (see Fig. 11), in particular, where traffic is heavy, there is a high possibility of near-misses or of being targeted by scrambled military aircraft because of straying from the air routes and violating national airspace. The degree of danger in those situations is thus evaluated as high. When flying in regions of light traffic or where there is radar monitoring by an air traffic control system, however, the evaluation of degree of danger is low.

In the case of taking off without clearance from air traffic control, also, if the visibility is good, there are no other aircraft or vehicles on the same runway, and no planes are coming from the opposite direction, then the degree of danger is low. If, however, as in the case of the Tenerife and O'Hare Airport accidents described above, the visibility is poor owing to dense fog or low clouds and there is another aircraft or vehicle moving on the same runway, the degree of danger is evaluated as 'serious', even if an accident was averted by the pilot making an evasive manoeuvre.

At airports that adopt the priority runway procedure system (in which aircraft that are landing and aircraft that are taking off face head-on), the degree of danger is high during the approach of the landing plane.

1.23 Breaks in the chain of events leading to an accident

The reason for recovery from an incident affects the evaluation of the degree of danger. For example, when overshooting altitude, the degree of danger is higher when the corrective manoeuvre is made as the result of advice or indication from another than when it is done as the result of the pilot's own recognition of the situation. That is because, when flying an airplane, an instant's delay in performing a corrective manoeuvre can by itself bring the situation close to becoming an accident.

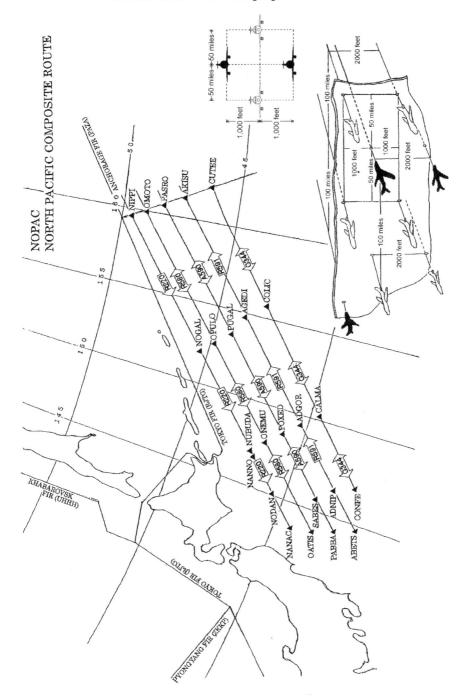

Fig. 11 North Pacific Composite Routes

In maintenance, too, for the same kind of flaw or problem, the degree of danger is higher when the flaw is missed in the service check and also not discovered in the preflight check by the pilot, but is discovered during flight after a problem occurs, as opposed to cases where the problem is noticed by the personnel that performed the work.

It is also rather obvious that problems affecting structural elements of the aircraft or emergencies are evaluated as having a higher degree of danger than those related only to comfort in the passenger cabin. Then, however, even for problems that relate only to comfort in the passenger cabin and have nothing to do with emergencies, concerned persons should take heed of the fact that, if the passengers notice the problem, it may cause the anxiety that some problem related to the airworthiness or safety of something they cannot see has been overlooked in the maintenance of the aircraft.

Concerning the work of the air traffic controller, too, the case in which the aircraft crew must make an evasive manoeuvre is evaluated as having a higher degree of danger than the case in which an air traffic controller attends to the abnormally small interval between two aircraft on the basis of distance (a 'confliction') or an abnormally close proximity (near miss), and issues instructions for evasion.

The evaluation of the degree of danger for reported incidents has the important significance of making it possible to discern from a macroscopic point of view the degrees of danger that are associated with the factors that have a possibility of constituting an accident, and that were ultimately found in information on incidents. A given danger factor may in one case constitute an unsafe event that has a high degree of danger, but in another case constitute an unsafe event that has a low degree of danger.

The danger factors that constitute a major accident do not necessarily all have high degrees of danger among the potential danger factors in that field. It is no exaggeration to say that nearly all of the individual danger factors that are discovered in incident information are the same as the factors that have constituted accidents in the past. However, improvements made as a result of advances in science and technology have changed the positions occupied by individual danger factors in relation to others, as well as the types of unsafe event that those factors constitute.

To keep up with that change, the incident investigations must be repeated at regular intervals. Through these repeated investigations, the present situation must be discerned: whether or not the improvement measures have been effective, what new kinds of unsafe events have occurred as the result of scientific and technological progress, and what kinds of danger factors now have high degrees of danger.   As is explained later, by using quantification method III to re-examine the results of analysing the information on individual incidents, the degrees of danger of individual factors among the many latent danger factors in that field can be clarified.

An incident is a case in which the chain of danger factors is broken for some reason (i.e. the reason for recovery) before it reaches the stage of an accident. Of the many incidents, there are some cases in which the flawed flow of information concludes without the operation of a reason for recovery or good fortune.

An example is the case in which an aircraft does not veer off the runway, but makes a hard landing that is not serious enough to be treated as an accident. Looking at maintenance work, there are cases in which parts break because of overtightening. That is to say, there are cases when some consequence occurs without the chain of events being broken by a reason for recovery, but it does not reach the stage of being a major accident, and such cases are treated as the occurrence of a consequence.

1.24 An overall picture of the danger factors

To ensure safety, it is of course desirable that appropriate improvement measures be taken for all of the latent danger factors that are discovered in incident information. This is because, as described above, the chain of latent danger factors is not necessarily rational, and it is possible for even a trivial error to link with other danger factors to cause a major accident.

In practice, however, we cannot expect to be able simultaneously to take measures against all of the latent danger factors found in incident information. We can only aim to lower the possibility of major accidents by addressing the many potential danger factors in a field in the order of highest degree of danger. On this point, the evaluation of the degree of danger of incidents is extremely significant for the prevention of accidents.

If it were not possible to clarify the degree of danger of the respective danger factors found in the information on various incidents from a comprehensive perspective, then the significance of searching for danger factors in incident information would be reduced by half. If an understanding from a macroscopic point of view were not possible, then we would be left with the question of whether our belief that some of the discovered factors have a high degree of danger is no more than hindsight after the accident has occurred. Whether before or after the occurrence of the accident, if the results are no more than an extension of the accident investigation, then the objectives of IRAS cannot be fully attained.

It is clear that incident information itself is important to some extent, but simple collection of a large amount of information (which is a very difficult task to begin with) and feedback of that information are not in themselves sufficiently useful in the prevention of accidents.

We mentioned earlier that the feedback of incident information, and specific incident information concerning the new types of danger that accompany automation in particular, is effective in preventing accidents. When MAIR is used to undertake an impartial analysis of the information on individual incidents, however, IRAS only becomes functional when improvement measures for the factors with a high degree of danger are further studied on the basis of a comprehensive multidimensional examination such as is described in the next section.

While MAIR also analyses the reasons why individual incidents ended without becoming accidents (i.e. the reasons for recovery), its significance is not limited to the evaluation of how close the incident came to becoming an accident, that is, the degree of danger. The various latent danger factors that are discovered in many incidents are classified according to each element for which there is a strong possibility of correlation. This is meaningful in that the recovery reasons that are effective for a class of unsafe events can be confirmed by noting which recovery factors appear in the same vicinity.

1.25 IRAS and quantification method III

The first step in preventing accidents is to obtain an objective understanding of the overall picture of the many danger factors that have been discovered in incident information and to clarify that overall picture.

In the past, various proposals for improvement have been made in the field of aviation as well as in other fields. However, there is a difference in awareness of the true state of affairs from the standpoint of the managers and administrators and from the standpoint of those at the actual scene of events. Moreover, it has not been possible to understand where the danger factors that have been identified from the results of investigating particular accidents or incidents stand in relation to the many other danger factors. There is no question that, as a result, the danger factors have not necessarily been linked to the correct improvement measures. To avoid the discord that results from different perceptions of the facts from this standpoint and to achieve effective improvements, the latent danger factors must be understood from a macroscopic perspective premised on objective understanding and clarification.

The feedback of specific information on individual incidents is important with respect to making it easier to avoid similar dangers that are encountered later. The IRAS that I have proposed, however, has important functions that go beyond this, to reveal the framework of the correlations among all of the factors that relate to the occurrence of the incident (i.e. that probably constitute accidents), and to show the factors with respect to which improvements should be made so as effectively to prevent accidents through an examination of the overall picture.

When we go no further than the simple analysis of individual incidents, the results are no more than a linear extension of the accident investigation, in that such results are useful only in the prevention of unsafe events that are similar or of the same type as the incident that was analysed. We are all familiar with the expression 'You can't see the forest for the trees'. In the same way as analysing an individual tree, no matter in what degree of detail, cannot give us an understanding of the entire forest, the organic living mass in which the individual tree lives, analysis of a single accident or incident, regardless of the degree of detail, cannot give us an understanding of the whole picture of the danger factors of a particular field.

As mentioned earlier, quantification method III, a statistical method of multidimensional analysis, is the most suitable tool for discerning the overall picture and framework. In 1956, Chikio Hayashi, Professor Emeritus and former Director of the Statistics Research Institute of the Ministry of Education, successfully employed this method to create a three-dimensional stereoscopic model of the elusive Japanese spirit, and it has also been used in many analyses of data-based phenomena in the romance language countries as well.

Professor Hayashi has said 'Numbers are a powerful means of exploring knowledge' (*Nihon Keizai Shinbun*, 19 October 1997), but quantification method III used in the IRAS is truly a powerful means of 'exploring danger'. Quantification method III is a method for quantifying the qualitative factors with the concept of the coefficient of correlation and clarifying the interrelationships among the factors. As suggested by the term 'multidimensional analysis', this method enables us to position the results of an overall analysis of various factors in a three-dimensional stereoscopic graph and so understand the strengths of the correlation among those factors (see Fig. 15 on page 122).

According to the theory of quantification method III, when looking at the $X–Y$ plane of the three-dimensional stereoscopic graph, factors that appear closer to each other have the strongest correlations and factors that are furthest from each other have the weakest correlations. Accordingly, we can say that factor groups that are distributed at the two extremes of the X axis have different characters, and, if that difference in character can be ascertained, it is possible to know two major characteristics of the danger factors.

The strengths of the factor correlations when looking at the stereoscopic graph from above (i.e. looking at the $X–Z$ plane) show next to the strengths of the correlations with respect to the factors as seen from the front (i.e. the $X–Y$ plane); the strengths of the factor correlations from the side (i.e. looking at the $Y–Z$ plane) show next to the strengths of the correlations with respect to the factors as seen from above (i.e. the $X–Z$ plane). The X, Y, and Z axes themselves have no quantitative meaning; they do nothing more than serve as scales for displaying the various factors that are computed by quantification method III.

1.26 Searching for incident patterns

If the factors that are positioned near to each other are grouped, taking the front view (the $X–Y$ plane) of the three-dimensional stereoscopic graph as reference, a number of typical incident patterns (referred to as clusters in statistics) can be discovered (see Fig. 13 on page 118).

Actual incidents do not necessarily involve all of the factors that make up each cluster; some may be missing or other factors may be included. It can be said, however, that the factors that constitute each cluster are the typical constituent factors of that type of incident. Accordingly, when considering the prevention of accidents, while it is of course effective to take measures against the factors associated with high degrees of danger, it is also effective to take measures against the main factors of each cluster.

The factors that are close to the 'serious' degree of danger represent a very high degree of danger; this can be seen from an overall view of how the various factors are related, obtained by taking an overall look at the danger factors of the field in question. Furthermore, it is possible to confirm what kinds of improvement are effective according to the incident type, by seeing which reasons for recovery are grouped nearby (explained in detail later in this book).

Through this kind of statistical processing, we can understand the kinds of interrelationships that exist among all of the factors found by MAIR to be related to an incident from the information on individual incidents and the overall image and framework.

1.27 Why take up the field of aviation?

It stands to reason that, because of the high speeds involved in the operation of aircraft used in commercial airline flights, regardless of the situation encountered, the aircraft cannot stop in midair to avoid danger or to consider how to avoid danger. This field furthermore differs greatly from other advanced technology industries in the possibility of encountering various weather conditions that have so far not been fully understood.

Even in the field of nuclear power generation, in which accidents can cause damage of the largest scale and most serious degree, it is possible to halt the operation of the reactor. Aircraft operating in commercial airspace face a greater severity than is encountered in other industries inasmuch as they cannot be stopped during flight.

While the aircraft itself is a large-scale complex system, the operation of commercial aircraft is supported by a variety of the most typical of large-scale complex systems, including airport and aviation safety facilities, air traffic control, and aircraft maintenance and operation. Although each of those support systems has its own safety systems implemented by an independent organization, one defect can affect others and ultimately have a large effect on the pilot's operation of the aircraft. Chapters 2 to 4 of this book describe the main latent danger factors that were found in incident information reported to the Japan Research Institute of Air Law by pilots, air traffic controllers, and aircraft maintenance personnel. At the end of each chapter, the results of an overall examination by means of quantification method III are presented.

What is described there is not limited to the single field of aviation, but is rather intended to provide a means of understanding and problem solving that is common to all fields of science and technology. In that sense, we take up the field of aviation as the issues it involves are common to leading-edge technology industries and complex large-scale systems.

1.28 Meaning of Dr Shigeo Okinaka's document 'Medical Record'

Dr Shigeo Okinaka, once known as a prominent doctor of internal medicine (the autonomous nervous system), in his final lecture before retiring from the University of Tokyo, demonstrated the difficulty of clinical medicine to the students by disclosing his own misdiagnosis rate, which had a great impact on society.[82] Moreover, in December 1984, 'Bitter charts – learning to prevent misdiagnoses from medical cases', a collection of the experiences that well known physicians have had with misdiagnoses, was published (Nikkei–McGraw-Hill), and, as one might expect, had a major impact on society as well.

Miyaki Mori, the editor of this book and chief editor for 'Nikkei Medical' at that time, said the following in the preface to that book:

Medical doctors are well aware of the inescapable fallibility that is the nature of medicine. Nevertheless, because 'failure is unacceptable' is a fundamental principle, physicians are less than willing for even their unavoidable mistakes to be brought into public view. That is probably all the more true when medical malpractice lawsuits are frequent, as they are nowadays.[83]

For this reason, medical failures are more often reported as simple statistical numbers or generalizations. One aspect of that, however, is that it interferes with the passing on of valuable experience.

It is with that in mind that Nikkei Medical established the 'bitter charts' column. Fortunately, many highly respected physicians have related their bitter experiences. The handing down of experiences in this way will provide input in the medical examinations of many doctors.

The significance of the sharing of experiences through the publication of incident information is exactly the same as the significance of 'bitter charts' to the field of medicine. The one difference, however, is that the authors of 'bitter charts' are all highly renowned medical practitioners. Because they are so highly regarded, their admissions in 'bitter charts' can be lauded by the public as brave acts, and past errors in diagnosis can be accepted with tolerance. In contrast, individuals that report incident information involving the human factors dealt with here in this book are unknown aviators, and the public evaluation of such 'bitter charts' reported by them would be severe.

However, in the not even one generation since the Wright brothers realized mankind's long-held dream of powered flight, aviation has undergone spectacular development to attain today's prosperous state of the industry. In this prosperity, the efforts of nameless aviators who have spotted even small problems and have been diligent in improving materials and technology must not be forgotten. Once those kinds of problem are understood in a systematic way in the form of IRAS and that content generally publicized, it often happens that this valuable information is played up in the mass media as 'failures that should not have occurred' or 'mistakes that should not have been made', stirring up public blame.

When the writer's fears are realized in that way, the reporters of incident information that involves human factors would probably not provide such information again so as to avoid public blame and personal detriment. Such a state of affairs goes against the purpose of IRAS, and is also extremely important with respect to ensuring safety, because the latent danger factors will remain hidden and the path leading to an accident will remain open. To avoid this sort of situation, and from the viewpoint of maintaining safety, it is necessary that the public have a thorough general understanding of IRAS. It should be realized that the manner of thinking that impedes that understanding is, itself, a matter of abandoning one's own safety.

The important thing is to derive the general probabilities of what kinds of error people make under what circumstances in a particular field, through the analysis of incident information that involves human factors, and then proposing specific countermeasures for each case. Of course, this in no way involves determining the responsibility of the writer of the report or other persons concerned, and there should be no investigation into the identity of the writer. From this perspective, one absolute condition in IRAS is the confidentiality of the identity of the writer of the incident information report, which is to say the source of the information.

In Chapters 2 to 4 of this book, we will see what kinds of danger factors were found from information on actual incidents that involved human factors, but items from which the identities of the reporting persons or other related persons could be deduced have been omitted. Also, expressions have been changed while remaining strictly faithful to the facts of the incidents and the descriptions in the reports.

There are, however, items whose omission would be inappropriate because they are essential to the report and for which inferences concerning the writers of the reports may be possible to some extent. It absolutely must be understood that making such inferences is not only counter to the purpose of IRAS, but is also detrimental to aviation safety.

Chapter 2

Pre-accident situations experienced by pilots

2.1 The work of a pilot

Commercial aircraft are operated by a cockpit crew of either two or three crew members including a captain, a copilot, and a flight engineer. The captain (pilot) holds a formal airline transport pilot rating (ATR) with the qualification of the specified type of aircraft. Also, the captain has been qualified by the same Ministry for the air route. The copilot holds a commercial pilot rating or higher. The flight engineer holds a formal certificate for flight engineering skills. All crew members are required to hold a certificate of flight medical examination issued by the Ministry of Transport. The operation of commercial aircraft under conditions other than those described above is not legally recognized.

In the cockpit of the aircraft, there are pilot's seats on the left and right, with the same instruments and switches arranged in more or less the same way on both sides. The captain usually sits in the seat on the left and the copilot is seated on the right. The flight engineer sits behind the copilot's seat facing the outside.

The captain is responsible for the operation of the aircraft. The copilot, while having about the same qualifications as the captain, generally has less experience than the captain. The flight engineer is a specialist concerning the mechanical systems and, in addition to being in charge of all of the aircraft's systems, including the engines, fuel system, oil pressure, and adjustment of the cabin pressure and temperature, the flight engineer usually handles communication with the passenger cabin. In the most recent 'high-tech' aircraft, such as the Boeing 737, DC 9 series, Boeing 747-400, MD-11, and Airbus 300, there is no flight engineer.

One of the pilots grasps the control wheel (pilot flying) and flies the plane, while the other pilot performs assisting tasks (pilot not flying) and handles communications, provides the aircraft operator with information required for aircraft operation, observes the situation outside the plane, and monitors the instruments. Whether the captain or the copilot flies the plane is decided by the captain on each occasion.

There are many types of instrument in the cockpit. With so many instruments to observe, the tendency would be to think that only two or three gauges must be watched, and the rest need be viewed only when necessary. In fact, the pilot scans the instrument panels, his eyes running over many gauges, and always checking the navigation instruments, including aircraft attitude, altitude, and heading, and the climb rate, speed, and engine status. Concerning the engines, the pilot must also keep an eye on the thrust and temperature, revolution speed, fuel flow, and so on. This visual acuity is the first to be diminished by fatigue[1] and ageing.[2]

In this section, I will describe the series of actions performed by the pilot from take-off to landing, to provide a general understanding of the pilot's duties.

When the pilot arrives at work, he first checks his own mailbox for the latest information such as air route revisions and inserts them in the manuals that he always carries. The printed material placed in that mailbox also often includes large quantities of labour-related notices from the company, campaign pamphlets, and so on.

After that, the pilot goes to the dispatch room to check in, check the weather for the destination and air routes to be travelled, and check the NOTAM (the notice to airmen, which includes flight information, notices to flight personnel concerning information on airport construction work that affects runways, parking aprons, runway lights and approach lights, and radio frequency changes; it also includes information about tall cranes in the vicinity of the airport, and so on.) When the captain, copilot, and flight engineer have been gathered together, the dispatcher conducts questions and answers, for example concerning the flight routes, flight time, and weather. The question session with the dispatcher also includes investigation of the weather report for an alternative airport to be used in the event that it is not possible to land at the destination airport.

The flight engineer checks the service record in the aircraft logbook in advance, and confirms any questionable points. Of particular interest are breakdowns that have occurred repeatedly even though maintenance has been performed. The causes of such failures are often not known with certainty, and the flight engineer may telephone the person in charge of maintenance to confirm items of interest.

There are often changes in the cockpit crew at flight time. The crew members have their respective rotation schedules and sometimes it is not known until the day of the flight who will be joining them in the cockpit. It is also common for the crew composition to change on account of illness and the need hastily to replace a highly experienced captain owing to minimum weather conditions according to pilot experience. Such sudden replacements are possible because of the standardization of jobs and the fact that all of the persons who perform the work have the same level and receive the same training.

The issue at this stage is the amount of fuel to be taken on. More fuel than is normally needed must be taken on because the lower engine thrust in places where the air temperature is high may require higher fuel consumption during take-off or because of the possible necessity of detouring to avoid typhoons or other adverse weather conditions.

This does not simply mean, however, that the more fuel carried, the better. Additional fuel increases the weight of the aircraft, which, in addition to affecting the climb rate and the required runway length, means that a lot of fuel is consumed just to carry the additional weight of the fuel itself. Therefore, finding an economical balance is a problem, and at times the position of the company and the opinion of the pilot are at odds.

2.2 Take-off and reject take-off procedures

The crew boards the plane 30–45 min before the scheduled departure time for domestic routes and 1 h before in the case of international routes. The pilot and flight engineer first inspect the exterior of the aircraft. Sometimes fuel leaks or other such problems are found at this stage. After entering the cockpit, they check the instruments and radio equipment, and, if there are no problems, sign a document accepting the aircraft from the maintenance officer.

The pilot explains the flight route, flight time, weather, and so on to the cabin attendants and receives a report from the cabin attendants concerning whether VIPs or ill persons are among the passengers. When the passenger cabin preparations have been completed in this way, the passengers are boarded.

The aircraft must be flown according to instructions from the air traffic controller in and near the airport and in the air routes. This means that the pilot must operate the plane entirely according to instruction from the air traffic controller, from the route for taxiing (moving on the ground) to take-off order, departure procedures, and the air route to use. Instructions from the air traffic controller include important information for avoiding collisions with other aircraft, so communication with air traffic control is one of the important duties of the cockpit crew.

The operation of the aircraft is generally divided between the captain and the copilot. The crew notifies the air traffic controller 5 min before starting the engines. When the air traffic controller has given the air route and altitude clearance (referred to as ATC clearance), the crew starts the engines while the aircraft is being pushed back from the gate by the towing car to a position from which the plane can move under its own power. When permission to taxi is received from the air traffic controller, the crew begins to taxi the plane.

While taxiing, the pilot checks the instruments, electrical power, cabin pressurization, fuel, and other systems, operates the flight control system, flaps, spoilers, brakes, and other equipment to test for normal functioning, and checks many other items, such as the brake and tyre temperatures. Moreover, the flight engineer (or, on aircraft that do not have a flight engineer on the crew, the pilot who is not in charge of flying the plane) reconfirms everything, reading off a checklist. Then a take-off briefing is held to discuss the take-off procedure and the measures, procedures, and division of duties in the event of an emergency.

When the preparation for take-off has been completed, the air traffic controller is informed that the aircraft is ready for take-off. When the air traffic controller grants permission to take off, he at the same time provides an information update, such as the current wind speed and direction. The pilot completes the final part of the pre-take-off checklist and then signals the passenger cabin that the plane is taking off by turning on the fasten seat belt signs and so on.

When given permission by the air traffic controller, the pilot moves the plane onto the runway, lines up the nose with the centre-line, increases the power to the engines, and begins the take-off. The pilot increases the aircraft speed while controlling its direction by steering with the handle that controls the nose wheels and by using the rudder. When the plane exceeds the speed referred to as V_1, the pilot removes his hand from the throttle control, even if one of the engines has failed, clearly demonstrating the intent to continue with the take-off even if a problem with an engine subsequently develops. If one engine fails during take-off, the take-off is aborted at speeds below V_1, but, in principle, if the speed is above V_1, the take-off is continued. Normally, for large aircraft, V_1 is from 150 to 180 mile/h. This lift-off speed depends on the weight of the aircraft, the runway surface conditions, the presence of rain or snow on the runway, and other such conditions.

When the speed at which the nose of the aircraft lifts up (V_R or 'rotation speed') is exceeded, the pilot gently begins to pull back on the control column. When the nose rises up and the wheels have lifted off the ground, the pilot flying the plane checks the instruments to confirm that the plane is climbing and then issues the 'gear up' (wheels up) instruction. At the same time, he performs a 'thumb-up' gesture with one hand to instruct that the landing gear be retracted. Then, climbing at the prescribed speed is begun.

In the case of reject take-off (RTO) due to engine failure or other such reason, the engine power is lowered to idle (the minimum thrust) and then the thrust is reversed or the propellers are braked. At the same time, the aircraft wheel brakes are applied and, to increase the effectiveness of the braking, a lever is pulled to raise the spoilers on the wings so as to increase air resistance and decrease the aerodynamic lift. These operations are performed by the two pilots according to the division of duties as confirmed in the take-off briefing.

As in the pre-arrangements made in the take-off briefing concerning procedures during take-off, various situations in the operation of the aircraft are continually hypothesized and anticipated. Accordingly, measures can be taken quickly to cope with the anticipated problems. On the other hand, extraordinary situations that cannot be anticipated can only be dealt with by coming up with a solution through the pilot's experience and spur of the moment judgement.

2.3 New emphasis on high-tech information-processing tasks

After the take-off has been completed, the plane flies by automatic control. In modern commercial airliners, automatic control has an important role. The automatic control equipment, connected to an Inertial Navigation System (INS) is capable of flying the plane automatically to the destination if the longitudes and latitudes of the departure point, intermediate points along the route, and the destination point are input. This system can also go as far as automatically beginning the descent and managing altitude. Accordingly, if the data input is incorrect, the plane cannot fly the correct course.

A pilot that has a relaxed attitude because the plane is under automatic control will not notice if the plane goes off course. Although the automatic control system is a convenient device that also displays wind direction and speed during flight, the ground speed, the current position, and other such data, if it does not function properly, it places a large burden on the pilot. As a result, it is the pilot's operation of the aircraft that must ultimately be relied upon.

With the introduction of more high-tech functions, aircraft systems have become more complex, and the pilot must monitor these complex systems to gain an understanding of the overall situation, from the weather and fuel supply to the condition of the passengers, and must make correct decisions. To be sure, operations such as throwing switches and moving the control wheel are now seldom performed while cruising, but such duties have been replaced with important work such as monitoring instruments, so the pilot cannot be lax.

The pilot must constantly pay attention to the sound of the engines. Because acceleration and deceleration can be felt with the body, the pilot is always paying attention to the vibration and force of acceleration transmitted to the body by the seat. To continue monitoring instruments for a long time while sitting still and maintaining concentration requires considerable physical strength.

Although it has been conjectured that the accident in which a Korean Airlines plane was shot down by a Soviet Sukhoi-15 in the sea off Sakhalin Island in September 1983[3] and the April 1978 accident in which a Korean Airlines plane was forced to land in Murmansk happened because the pilot was asleep, pilots do not normally sleep at the controls.[4] There is, however, a phenomenon called 'microsleep'[5] that can occur during long-distance flights of 10 h or more. In this phenomenon, the pilot falls asleep for extremely short periods of time but is unaware that this is happening.

The captain also bears ultimate responsibility[6] for disturbances caused by passengers in the passenger cabin and hijackings, and must deal with such situations. On international flights, the captain can order the restraint of any person guilty of committing a crime or other of specified behaviours aboard the aircraft, and can have such individuals deplaned at the first country of landing and handed over to the relevant authorities.[7] In such an event, the decision is the captain's responsibility.

2.4 Landings require extra care

When the aircraft is approaching the destination it comes time to land, but the preparation for landing begins quite some time before that, with notes being taken concerning the weather conditions at the destination broadcast by shortwave or UHF radio. It is no exaggeration to say that thinking about the landing starts when the plane begins level flight. Tentative plans are made for various contingencies, from the direction of approach to the runway depending on wind direction to which airport to divert to (destination change) in the event that the destination airport is closed because of weather, accident, or other unexpected conditions.

In order to land at Haneda International Airport near Tokyo, for example, the pilot then prepares in his mind many flight decisions, such as how many nautical miles before reaching Oshima Island to begin descent considering the effects of the plane's altitude and the wind, at what point to change to level flight for the purpose of reducing speed, and at what time to lower the flaps. These decisions extend even to where to lower the landing gear, which taxiway to use to get to the designated parking spot after touching down, and whether hard braking is needed for that purpose.

Before landing, a landing briefing is held by the cockpit crew, in which prearrangements are made for a missed approach or go-around. An approach retry occurs when landing is not possible because the runway or approach lights are not visible, in which case the approach is abandoned and the plane is made to climb; a landing retry is when the runway is visible, but landing is not possible.

When the descent is made and the stage of actual landing is reached, the landing gear is extended, the approach line is correctly lined up with the Instrument Landing System (ILS), and the speed of the plane is reduced. If it is decided that landing is not possible, full power is applied to the engines and a missed approach or go-around is executed.

If the runway is discernible, the pilot pulls on the control column when the plane is about 33 ft above the ground and power to the engines is reduced. After touchdown, reverse thrust is applied immediately. The spoilers are raised automatically when the thrust is reversed. The wheel brakes are also applied automatically on many types of aircraft, but, when automatic braking is not used, the pilot applies the wheel brakes to slow down the plane. When the plane has slowed down to about 80 knots (about 84 mile/h), the reverse thrust is stopped and the speed is reduced further. At the point for exiting the runway, the spoilers are retracted, the flaps are raised, and the plane is headed for the parking spot.

While the plane is moving on the taxiway, the landing time is reported to the company and the parking spot is confirmed. During this time, the Auxiliary Power Unit (APU) is started up, and the taxiway is followed to the parking spot. After the plane has stopped, the power to the engines is cut off and the seat belt signs in the passenger cabin are turned off. After entering the parking spot, the crew runs through the parking checklist, confirming the landing time, time of arrival at the parking spot, and other information. Any problems that occurred during the flight are recorded in the logbook and the log is signed and handed over to the maintenance officer. The pilot then goes to the dispatch room to end the flight duty period.

In this way, the pilot's work is mainly a matter of collecting information and making situational decisions, and, on that basis, operating the aircraft and anticipating and preparing for situations as described above. However, even with the intent to anticipate various situations and to know the latest correct information, misreading or overlooking of instrument displays may lead to easy-going/assumption decisions or wrong convictions about the situation and on rare occasions to difficulties in operating the aircraft.

Aircraft fly in weather conditions that change from minute to minute. Even for veteran pilots, it is safe to say that no two flights ever present the same conditions. Surprisingly, many pilots and flight engineers have had frightening experiences in which, by good fortune, an in-flight situation did not result in an accident because of advice from other crew members or air traffic control, the operation of warning devices, or because the pilot himself noticed a problem and performed a corrective operation.

2.5 Fatigue as a major factor in errors

We cannot effectively ensure safety unless we identify the background factors that lead to human errors and devise measures for them rather than focusing on the errors themselves, regardless of how severely the persons making the errors are treated. In identifying those background factors, it is necessary to search for them in information on specific individual incidents as described in Chapter 1 rather than hypothesize them in advance.

Here, however, I reverse the actual process of analysis for convenience in explanation, focusing on what kinds of background factor there are among those discovered in 430 incidents that are related to aircraft operation. In the following, I shall attempt to show the kinds of background factor included in the information flow to induce pilot errors, and the degrees of danger and the reasons for recovery in those incidents.

The approach to an airport on the continental United States was made with the reading of the barometric altimeter set high by about 1 in (about

1000 ft difference in altitude). Because such low atmospheric pressure is seldom experienced, the captain and copilot both believed that the reading was correct. It was known that the crew's level of attention was clearly lowered due to fatigue from a long flight that involved jet lag.

The thought of what would have happened if we had been flying through clouds in a non-precision approach (in which information on the angle of descent is not obtained by radio) sent a chill down the captain's spine. We were saved because the flight engineer noticed the problem.

Regardless of the fact that the atmospheric pressure was abnormally low, the barometer was set to a pressure near the normal value (an incorrect operation) according to an easy-going decision based on a preconception, and the approach was continued. The background factors were a conflicting combination of fatigue due to the long overnight flight and the time-zone difference (jet lag), the abnormal atmospheric pressure, and familiarity. The stage in the information process that determined the incident was the decision phase. The degree of danger was 'considerable' and the reason for recovery was advice from the flight engineer.

The aircraft altimeter employs the relationship that the atmospheric pressure is lower at higher altitudes, calculating the altitude from the measured barometric pressure and displaying that value. Because the atmospheric pressure at ground level changes with the weather conditions, if the atmospheric pressure is particularly low, the correct altitude above sea level will not be displayed unless a corresponding compensation is made. For that purpose, the ground-level atmospheric pressure (called the QNH) is input as a compensation device.

The evaluation of the degree of danger associated with incorrect setting of the pressure altimeter differs greatly according to the specific circumstances, including: the flight phase (on the ground, take-off, climbing, cruising, descending, approach and landing); the difference between the correct atmospheric pressure value that should have been input and the incorrect value that was input; the time at which the error was noticed; and the presence or absence of other aircraft in the vicinity.

The greatest danger posed by error in setting during landing is when the atmospheric pressure is set higher than the actual value, as it was in the example incident just described. Although the pilot thinks that the altitude is as displayed by the altimeter, the actual altitude is less than that. Therefore, if a non-precision approach is being made in clouds so that no information is available from outside the plane and the erroneous setting is not noticed, there is a danger of the plane hitting an obstacle on the ground or hitting the ground surface.

There are examples in which fatigue and heavy workload combined with abnormal atmospheric pressure so that the wrong atmospheric pressure was set because the pilot did not correctly hear the adjustment value spoken by the air traffic controller or with post-stress carelessness during climb so that the pilot forgot to set the readjustment value to QNE (standard atmospheric pressure, 1013.2 hPa or 29.92 inHg).

The reasons for recovery in these cases are most often the checking function of the crew or advice from air traffic control. There are, however, examples in which mistaken information from the air traffic controller poses a 'considerable' degree of danger, when the pilot notices during an ILS approach that the altitude above the outer marker is higher than the value entered in the approach chart. The outer marker is one of the facilities that constitute the ILS. It is usually installed at a point 5 nautical miles from the airport and transmits a directional radio beam upwards. For a precise ILS approach, the altitude that should be displayed by the altimeter on the aircraft above the outer marker is specified so as to prevent the plane from following an incorrect glide slope (a radio wave course that shows the angle of approach and landing). The direction radio beam of the outer marker makes it possible to know the position of the aircraft flying over it with a high degree of accuracy.

There are many examples of unsafe events, including accidents, for which fatigue due to long-distance flights and jet lag is a background factor. Fatigue lowers a person's level of awareness and makes one susceptible to failure to notice and excessive focus (in which one cannot divide one's attention) and failure to see, as well as delay in information receiving, making decisions, and taking action or giving instructions, easy-going/assumption decisions, and incorrect or inappropriate actions/instructions.

There follows an example of a 'considerable' degree of danger in which the aircraft almost overshot the runway. After 9 h of flying, the aircraft was making the final approach to the destination airport. The aircraft was above the glide path, but the pilot delayed adjusting the rate of descent. When the pilot took corrective action, the aircraft was relatively high above the glide path. The pilot therefore increased the rate of descent, which also increased the airspeed to about 30 knots (33 mile/h) above the prescribed speed. The pilot extended the flaps to 30° to reduce airspeed, but airspeed was too high for 30° flaps. The flap protection mechanism was activated to limit the flap deployment to 25°, so the airspeed did not decrease as expected. Although the flaps finally extended to 30° at the threshold (the approach end of the runway), the aircraft had landed at 4000–5000 ft beyond the threshold. The normal touchdown point is 1500 ft from the threshold. This was an example of recklessness and forceful operation of the aircraft.

Other than this, there are examples of 'virtually none' degree of danger incidents in which the pilot mistakes a switch for a similar one because of fatigue due to jet lag, but immediately notices the mistake and corrects it. Then there are examples of 'fairly little' degree of danger incidents in which the pilot absent-mindedly forgot the altitude restrictions because of fatigue due to lack of sleep, and was advised by an air traffic controller just before exceeding the altitude limit.

When fatigue combines with other factors such as weather, unexpected changes, diversion, runway change, interdependency, and fluster and impatience, the tendency is for the degree of danger to increase. Factors concerning the pilot's physical condition other than fatigue that are found in the reported incidents include a number of cases of sudden illness while on duty, and one case each of the effects of alcohol consumption and drugs.

2.6 Impatience, fluster, and carelessness after stressful periods

The presence of low clouds which can change visibility at any moment and awareness of restrictions on working time imposed by the company can lead to the mental state of impatience. The inference that the desire to take off quickly has contributed to causing accidents has been described earlier in this book, concerning the KLM pilots in the Tenerife accident.

It happened on the fourth flight of that day. Because it was the last flight, there was probably impatience from a strong desire to return quickly. After setting the engines to take-off power while making the take-off run, the acceleration was unbearably slow, so the pilot commented to the copilot that 'The acceleration is poor'. Then, the copilot noticed that the water injection switch (a device in the engine that injects a solution of water and alcohol to lower the temperature and increase thrust) had not been turned on, and hastily turned it on.

Turning that switch on while the engine is at maximum output may damage the engine, and because that is very dangerous, the take-off was cancelled.

As a result, the water injection was not activated, but, if it had been, it would have been dangerous. The reason the incident occurred is that, in spite of the inspection according to the checklist, the copilot had responded that the two green light indicators for the water injection were on without actually confirming that they were in fact on.

This example involves a chain of three events: the copilot forgetting to turn on the water injection switch (the first event), failure to notice that the switch was not turned on when going through the checklist and wrong conviction leading to the answer that it was 'on' (the second event), and the easy-going decision to turn on the switch upon hearing the captain's comment that 'The acceleration is poor', even though the engines were at maximum output (the third event). In addition to impatience and fluster, this incident involved the background factors of fatigue – failure to see and responding that the indicator lights were on out of habit rather than actually verifying the fact by sight when going through the checklist. The degree of danger is 'considerable', and what determines the incident is the third event; that is, the easy-going decision.

The following is another example in which impatience or fluster is a factor in the same way. After landing, the pilot turned the plane across the runway towards the parking apron on instruction from the air traffic controller. Another small aircraft that was moving in front of the airliner suddenly stopped, so the pilot of the airliner was also forced to stop momentarily. The tail of the airliner, however, was protruding out into the runway, creating the possibility of being hit by other aircraft that were taking off and landing. The pilot therefore excessively focused his attention on the movement of the small plane in front of him, fixating on moving past it on the right side. The pilot's actions were so reckless, forceful, and excessive that, although the tail of the aircraft was no longer protruding onto the runway, the aircraft nearly ran off the taxiway. The degree of danger associated with this incident is 'serious'.

Other pilot psychological factors found in reported examples include stress, inattentiveness (post-stress carelessness), tedium, and uneasiness or worry. It goes without saying that, whatever the job, an appropriate degree of stress while working is desirable, but people cannot keep up such states of tension constantly. There are cases in which excessive stress may bring about psychological states of impatience and fluster depending on the situation, or, conversely, post-stress inattentiveness (carelessness) or tedium, which may lead to errors.

When operating an aircraft, in particular, stress is high at times of unexpected danger such as near-miss situations with other aircraft and sudden deterioration in weather conditions, and during pilot training or check flight take-offs and landings. On the other hand, it is easy to fall into a state of carelessness after relief from a highly stressful state. There are also cases in which the inability to maintain the required level of stress owing to fatigue puts a person into a distracted daze.

Even though pilot impatience or fluster in response to a close call with unexpected danger can induce errors, it rarely happens that excessive stress leads to errors. Rather, the state of carelessness that follows after dealing with that danger often does lead to errors. There have been reports of 'virtually none' degree of danger incidents upon entering closed taxiways after the stressful landing operation has been completed and the crew is in a relaxed state of mind; this is one such example.

Nearly all examples in which stress is a background factor that leads to errors occur during pilot training or check flights.

If smooth flight continues over a long time under automatic control, the crew members may converse about matters unrelated to the job, seeking suitable stimulation to relieve the tedium in the small cockpit, which in some cases leads to errors of 'failure to notice' or 'forgetting'. In an example of the 'virtually none' degree of danger, workload was light because of good weather conditions, and, while the plane was climbing at an altitude of 14 000 ft, the entire crew forgot to switch the adjusted value of the altimeter to the standard atmospheric pressure (QNE) because they were conversing. This caused an error of about 200 ft in the altimeter reading, but the crew corrected the error on advice from air traffic control.

2.7 Mistakes caused by much or little experience and knowledge

The personal factors of crew members that are found to be background factors in unsafe events are degree of experience and knowledge. It is easy to think that having a great deal of flying experience normally has a positive effect on safety. That is not necessarily so, however, because having abundant experience can be a factor in giving rise to errors that result from habits.

There are surprisingly many examples in which habit is a background factor in inappropriate or mistaken operations based on insufficient checking or easy-going decisions due to inattention. Concerning the checklist inspections that are made several times in the course of every flight, in particular, there are occasional cases in which, even though the eyes are looking in the direction of the instruments, the responses to the checklist items are made habitually, without actually confirming the indicator values or switch positions.

One example of an incident of 'considerable' degree of danger is failure to see an incorrect take-off thrust setting because the three crew members simply read the checklist at the time of the pre-take-off checklist inspection. If the take-off thrust is set too high, the engine may be damaged; if the thrust is set too low, it will not be possible to take off at the planned runway length. In the background of this example is the interdependency of the crew members.

In another example where the degree of danger was 'fairly little', a captain who had nearly 10 000 h of flying experience, but was flying this particular air route for the first time, was landing at an airport whose runway was 4000 ft long and 150 ft wide. Because he was used to seeing runways that were 4000 ft long and 100 ft wide, the mistaken conviction he had when viewing the trapezoidal shape of the runway from the plane above after beginning the approach was that the runway was 5000 ft long, with the result that the plane almost overshot the runway. In fact, the plane came to a stop 650 ft short of the end of the runway.

When a person reacts to a given situation, particularly when the person does not recognize that the situation differs from the person's general past experience, the tendency is to base the decision on the past experience. Basing an inference on past experience in spite of the difference with the present reality leads to error in the understanding of the situation (wrong conviction).

Insufficient experience and unfamiliarity are background factors that lead to vacillation and delay in the information receiving and decision phases, and to operation that is unreasonable or high-handed, or difficult, incorrect, or inappropriate.

There are examples of serious degree of danger incidents with background factors leading to easy-going decisions or difficulty in operation that involve runway overruns that occurred because the wind shear phenomenon was unknown to the aviation industry at the time, as well as serious degree of danger incidents in which little knowledge of the effects of snow and ice build-up on the wings on lift and acceleration was a contributing factor. There are also examples in which the pilot had no knowledge or experience concerning how the runway can be seen when the ground is blanketed with snow, and was thrown into vacillation in the decision phase.

2.8 Danger arising from interpersonal relationships among crew

The operation of commercial airliners involves multiple flight crew members. The link between failure to maintain smooth relations among those crew members (crew co-ordination) and unsafe events has been demonstrated by the relation of the captain and copilot of the KLM plane in the Tenerife accident, and by the relation between the Japanese Air Self-Defence Forces captain and the copilot in the serious ground collision incident at Naha Airport. The copilot in the latter case was undergoing training drills for rescue operations in the MU-2 aircraft.

For smooth intercrew relations, it is necessary to have a clear division of duties and a common understanding of the situation, regardless of differences in the experience of the individuals. If there is a shared understanding with a clear division of duties, even if one crew member commits an error, danger can be averted by the advice or action of another crew member. If the division of duties is inappropriate or unclear, however, there is a possibility of falling into a state of danger because the capability for correcting one error is not in effect or the entire crew may have its attention focused on a single problem.

There follow a few examples from among the 430 incidents related to the operation of the aircraft in which interrelations of the crew members were one of the factors.

When there is a feeling of interdependency among the members of the crew in the cockpit and no clear division of duties, the crew members unconsciously fall into a mental state of dependence on the other crew members (interdependency), in which it is believed that somehow someone else will perform an action or notice a problem.

The previously described case, in which habit formed by abundant experience led to the failure to see a take-off thrust setting that should have been discovered during the checklist inspection, is one such example. There are also cases in which the meaning of a person's operation or instruction was not clearly conveyed, but there was a mistaken conviction that it was understood and confirmation was neglected.

In addition to interdependency, there are other factors that are related to the interpersonal relations of the crew members. Although a relationship of mutual reliance is necessary when several persons co-operate to accomplish a single task, if there is excessive confidence in the judgement or actions of the captain or other highly experienced crew members (excessive reliance), then a person may hesitate to offer advice or be unable to receive advice from others altogether.

There was an incident whose degree of danger was 'virtually none' in which the captain mistook the assigned altitude cleared by the air traffic controller for the altitude requested by the aircraft in the pre-take-off briefing. The other crew members failed to notice the mistake because of habit, interdependency, and excessive reliance on the captain, so the plane exceeded the assigned altitude by 500 ft, which was then corrected on advice from air traffic control.

Another case of 'serious' degree of danger involved a veteran pilot who was receiving training for missed approach procedures under the hypothetical condition of No. 1 engine inoperative. Because of the trainer's excessive confidence in the experienced pilot, he was relaxed and his seat was pushed back further than usual. The trainee had neglected to increase engine thrust, which is an essential operation for go-around, because his attention was concentrated on pressing the go-around mode switch of the flight director (a palm switch designed to be slightly difficult to push so as to prevent erroneous operation). The trainer was slow in taking the corrective action, however, because of his relaxed position.

There are also cases that involved background factors such as questions not being asked because of excessive awareness of the other person's point of view, or advice not being given owing to misplaced deference or excessive concern for the other person.

An example of this is an incident of 'considerable' degree of danger in which a highly experienced copilot was working with a captain who was not accustomed to the type of aircraft he was piloting. While on the approach to the runway in a strong crosswind and rain which was above the prescribed landing weather minimum, the copilot, showing excessive concern for the less experienced captain ('I thought it might give the impression of belittling the captain'), refrained from offering the advice to go-around. As a result, the plane made a hard landing.

In another such case, the flight engineer noticed an error in operation by the captain, who was not familiar with this particular route, but waited for the copilot to notice it, thinking 'It's better for the copilot to advise the captain than for me to do so'. It was not until the aircraft neared the stalling speed that the flight engineer spoke up. In this case, the altitude of the plane was sufficient to allow recovery, and, because it did not reach the stage of buffeting (a vibration caused by separation of the airflow on the upper surface of the wings), the degree of danger was 'fairly little'.

The captain must exert an appropriate degree of leadership in the cockpit. Without sufficient leadership, there will not be a clear division of duties among the crew members or co-ordination of their work. If, on the other hand, there is excessive leadership and meddlesomeness in the work of others, then it will be difficult to get advice from the other crew members, and the ability of the other crew members to make decisions will be reduced.

There is an example of an incident of 'considerable' degree of danger in which the copilot's judgement was impaired during training for promotion to the rank of captain because the trainer had scolded the copilot vehemently and the copilot entered a runway that was in use without clearance from air traffic control. Other than that case, however, most of the incidents for which meddlesomeness was a background factor occurred during pilot training or check flights.

Among those, however, there are also cases in which unsafe events occurred with lack of leadership by the captain and misleading on the part of the excessively self-confident copilot as background factors.

Meddlesomeness by an aggressive copilot, who was paired with a captain who was passive by nature, is said to have contributed to an accident in which the French Air Inter flight 5148 Airbus 320-111 aircraft crashed and burned on approach to Strasbourg Airport on 20 January 1992, killing 87 passengers and crew members and seriously injuring eight passengers and one cabin attendant.[8]

There is also a case in which the other crew members lost confidence in the captain as the result of his attitude while on duty, a situation that developed further into a feeling that the captain was unreliable. This created a grave atmosphere in the cockpit that became an impediment in later flights.

2.9 The dangers of the missed approach

It was early morning, before dawn. After a flight of more than 8 h over ocean, we were near the destination airport, but the weather conditions at that airport were below the minimum, so we were in a holding pattern, unable to land. Fuel was low, and unless the weather improved within 10–20 min, it would have been necessary to divert to land at another airport.

When I thought that it was the last holding circle that we could make, we received a report from air traffic control that 'Weather conditions above landing minimum'.

Although we hastily began the approach, air traffic control instructed successively lower altitudes. We hurriedly went through the checklist confirmations, but, when I happened to look forward, the plane had dropped 700 ft below the permitted altitude and was still dropping.

Agitated, I advised the captain that 'Our altitude is 3000 ft' and the captain immediately put the aircraft into a climb.

After that, following the instructions from air traffic control, we landed without problem. We were tired after the long flight that day, and impatient because of the low amount of remaining fuel.

The notification that the weather had improved sufficiently to allow the landing approach (i.e. better than the minimum weather conditions) was received within the holding time based on the amount of fuel required to reach the alternate airport and the approach was begun, but the altitude check and maintaining the appropriate altitude during the approach were forgotten. The background factors were the below-minimum weather conditions, the low remaining fuel, the heavy workload, the long flight, fatigue, and impatience and fluster. The degree of danger was 'considerable'. Development into an accident was averted because of the copilot's advice.

Each airline company sets its own standards for minimum weather conditions according to the standard specified in the Annex of the Convention on International Civil Aviation for the purpose of ensuring safe aircraft take-offs and landings. The minimum weather conditions for take-off and landing are determined according to the aviation safety facilities, the approach method for each airport runway, and the pilot's experience. In the case of the Tenerife accident described earlier, the minimum weather conditions for take-off included a runway visibility of 2600 ft, and, although the visibility was below that, the KLM plane began to take off.

If the minimum weather conditions are not satisfied, a plane must not take off or land. If the weather at the destination airport does not satisfy the minimum weather conditions, aircraft must be placed in a holding pattern to wait until the conditions improve, or the destination must be changed to another airport. If the weather conditions change so that the minimum conditions are not satisfied during the approach, the situation is in principle treated as a missed approach (the landing is aborted).

There are cases in which the pilot lowers the nose of the plane more than is necessary in an attempt clearly to see a runway that can be seen only indistinctly under the very minimum weather conditions for landing. The attitude of the plane in such cases is referred to as 'duck under',[9] because it brings to mind the posture of a duck when it thrusts its head under water. There are many examples of accidents in the past that involved hard landings or the aircraft hitting approach lights and so on, which can easily occur when there is fog or other impediments to visibility.

Even if minimum weather conditions are satisfied according to official observation data, however, the weather occasionally changes from moment to moment, so, when it is decided that the actual visibility is worsening, the situation should obviously be considered to be a missed approach. Therefore, when the official observation is that the weather conditions are the minimum conditions or better, it does not necessarily mean that the plane must land.

2.10 Heavy rain causing poor visibility

We were making the approach at night under the Instrument Landing System (ILS). The weather was deteriorating and there were intermittent thunderstorms. When we began the descent, the thunderstorms briefly abated and we received a radio communication from the company that visibility had also improved. At the same time, we also received information that the next thunderstorm was approaching the airport.

Flying under ILS, the runway, approach lights, and so on could be seen clearly until we reached an altitude of 500 ft. Suddenly, however, we entered a cloud and, to make matters worse, heavy rain. We lost all visual cues and the sound of the violent rain obscured the call of 'minimum' (the altitude during the approach at which it must be decided to make a missed approach, when the runway or visual cues on the ground cannot be discerned visually). It was all the captain could do desperately to hold on to the control wheel and the copilot, too, was in no position to advise the captain to make go-around. The aircraft made a hard landing on the pitch-dark runway.

Heavy rain impairs visibility, and the forward visual range is reduced at the time of take-off or landing. In the example just described, visual cues were lost because the aircraft suddenly entered cloud and heavy rain at an altitude below 500 ft, and the 'could not see' factor in the information receiving phase led to unreasonable decisions and reckless, forceful operation. Although there was no injury to persons on board and no damage to the aircraft, the three crew members were described as having received such a strong shock that they said nothing from the time the plane entered the parking apron until after they had arrived at their hotel rooms.

Because the destination airport was situated in a location surrounded by mountains, the pilot had a potentially strong inclination to be reluctant to make a missed approach. If the required visual cues (the runway, approach lights, etc.) are not visible at the decision altitude, then naturally a missed approach should be made, but it can be inferred that the opportunity to make a missed approach decision was lost because the captain was concentrating on stabilizing the plane and the sound of the violent rainfall obscured the 'minimum' call. That the incident did not develop into an accident is nothing more than 'good fortune'. The degree of danger was 'serious'.

Roughly 5 years after that incident, on 12 May 1986, a Japan Airlines Boeing 747 was landing at the Hong Kong Kai Tak International Airport and there were no problems at all in the approach. However, when the plane passed the threshold of runway 13, and just as the engines were throttled down for touchdown, heavy rain was encountered. Forward visibility was sharply reduced, the speed of the plane was slowed by 10 knots (about 11.5 mile/h) in an instant, as if it had plunged into a pool, and the aerodynamic lift was reduced. As a result, it touched down 500 ft too soon.[10]

In another incident, when a plane was taking off from runway 18 of Naha Airport, cumulonimbi were approaching in the vicinity of the airport, but, with no special information received (the lack of weather information is a conflicting factor), the plane proceeded to take off. Soon, heavy rain struck and forward visibility was lost. The reason for recovery in this incident was the 'good fortune' that 'they were saved by the fact that the take-off path was over the ocean so that there were no obstacles'. The degree of danger was 'fairly little'.

Also, there was an accident on 27 July 1970 in which a Flying Tiger Line air freight flight DC-8 making a ground-controlled approach (a radar-guided landing in which air traffic control directs the approach according to ground radar) to runway 18 of Naha Airport crashed into the sea 2200 ft before the end of the runway.

According to the US National Transportation Safety Board, which investigated that accident, the characteristic strong sunshine and blue skies were present in the south, but scattered cumulonimbus and locally strong rainfall and wind shear had been expected. It is inferred that visibility was less than 1 mile when flying through the rain, but, when the plane passed through the wall of rain, it came out into the sunshine where it was 10–100 times brighter. It was inferred that the sudden change in brightness momentarily blinded the pilot, who was then unable to see anything and so the plane crashed.[11]

The sudden downpour of heavy rain is a characteristic weather phenomenon of the Southeast Asian region, and, because the area of rainfall changes from moment to moment, timely prediction by means of the usual periodic observations is all but impossible. In actuality, the pilot obtains the information from the plane that has just landed or taken off and from the air traffic controller or via the airline company radio. Sometimes, however, information from air traffic control or the airline company radio is not available (lack of weather information), and many unsafe events are caused by heavy rainfall in the vicinity of the airports of this region.

2.11 Danger caused by snow

Snow is another background factor that greatly affects the pilot's vision and operation of the aircraft. Snow can lead to the errors of 'could not see', 'could not see clearly', or 'saw incorrectly' in the information receiving phase. When snow accumulates to cover the ground completely, the visual cues that can normally be seen on the ground cannot be seen, and the visual impression becomes completely different. Against the solid white snow-covered ground, the white lights of the VASI (visual approach slope indicator) should be visible when the aircraft is flying at or above the approach angle. With the VASI, the pilot sees white lights when the plane is above the approach angle and red lights when the plane is below that angle. The white light is diffused by passing through the snow, however, and a reddish light can also be seen, so from the plane it is sometimes difficult to distinguish whether one is seeing the red lights or the white lights.

There is an incident in which all three crew members were vacillating in their decision of 'What light is that?' They continued to look for the runway and, finally, seeing other planes waiting on the taxiway, realized that their own plane was near the end of the runway. Because their altitude was high enough, they were able to execute a go-around and the degree of danger was 'virtually none'.

There is also an incident of 'considerable' degree of danger in which the touchdown marker was not visible and, with unreasonable decision and recklessness or forceful operation (or instruction), a plunging landing was made at excessive altitude and speed. In another case, the taxiway markers could not be seen because of accumulating snow at night, and the pilot turned onto a runway that was in use (active runway), thinking that it was the taxiway, and making it necessary for another plane to execute a landing go-around. The degree of danger in this case was also 'considerable'.

In yet another case, a plane was landing at a common-use military and civilian airport that had no runway centre-line lights after a 20 min wait in the air because of snow. Because snow was accumulating and the pilot could not see the centre-line, and because there were no other identifying objects, when the runway became visible he took a row of lights that could be seen in front of him to be the runway centre-line lights, although actually they were the runway lights on one side of the runway, and flew the plane directly over them. On advice from the copilot, the pilot executed a go-around. The degree of danger in this incident was 'serious'.

Even when the runway does have centre-line lights, they can hardly be seen when snow accumulates to about 2 in or so, unless snow is removed. The reporter of this incident information wrote: 'Braking effectiveness (on the runway) should have been tested and the snow should have been removed from the centre lights'.

In another case in which the degree of danger was 'serious', the crew could see almost nothing outside the cockpit on account of heavy snow during a daytime approach. The crew had executed missed approach 2 times by circling to the right, during which time it was difficult to see the runway from the captain's seat, and, just when the pilot was throttling up for the third missed approach, they caught a glimpse of the ground. Both of the pilots were excessively focused on finding the runway and the nose of the aircraft dropped excessively. When the ground proximity warning device sounded, the plane was at the dangerous altitude of 250 ft and continuing to descend at an angle of 4–5°.

The strong reflection of light by snow increases the time it takes the eye to adjust to relative darkness, so straining to see something on the snow-covered ground makes it difficult to see the instruments in the cockpit. While trying to spot ground markers according to information from outside the aircraft during a circling approach, the pilot missed sight of the runway. After about 8 min of straining to find the runway over the snow-covered ground, continuing the circling approach in the opposite direction, the pilot turned his eyes to the instruments to execute the missed approach. Unable to see clearly, however, he managed to execute the missed approach relying only on the instruments he could barely see. This is an example of an incident in which both the decision and the operation were appropriate, but in which the degree of danger was 'considerable'.

There is also an example of a 'considerable' degree of danger incident in which snow reduced the coefficient of friction of the runway or taxiway surface and thus lowered the braking effectiveness, making it difficult to manoeuvre the plane and resulting in the plane going off the runway. Landing clearance was given even though the runway had been closed 2 or 3 min earlier, because the air traffic controller showed excessive concern for the fact that the plane had already entered the final approach course of the traffic pattern (see Fig. 21 on page 152). However, no information on the braking effectiveness on the runway was provided, so the lack of information in advance and inappropriate instruction from air traffic control were background factors.

Low braking effectiveness conditions, unlike low visibility, cannot be known by observation from the pilot's side. It is not until touchdown that the pilot can be aware of the braking conditions. In Japan, moisture on the runway surface is divided into three levels, or, if snow and ice are present, six levels according to observations by the airport management. Those results are reported to the pilot of the landing plane via the dispatcher of the airline corporation (the operations manager).

In an accident that occurred on 16 December 1975, the Japan Airlines flight 422 Boeing 747 was proceeding on an icy taxiway for take-off from the Anchorage International Airport when the plane began to slip owing to a strong crosswind and slid backwards down the slope at the side of the taxiway, seriously injuring two of the 121 persons on board. Even though the airport manager had anticipated icy conditions, the fact that the improvement measures taken were slow and insufficient has been pointed out as a contributing factor.[12]

2.12 Snow and ice build-up on the plane as a factor posing the most danger

The most dangerous unsafe event in which snow is a background factor is snow and ice build-up on the plane, and particularly on the wings. If there is a slight amount of snow and ice build-up on the wing surfaces, buffeting (a vibration that occurs prior to a stall) will occur at a higher speed than usual, and there is a danger that it may lead to stalling of the aircraft. If the angle of attack of the wing is increased, the aerodynamic lift is increased proportionately. If a certain limit is exceeded, however, the airflow over the wing begins to separate from the wing surface and buffeting begins. If the angle of attack is increased further, the aerodynamic lift decreases sharply, at which time the aircraft stalls.

The speed at which the stall occurs is referred to as the stall speed. The lower stall speed is less dangerous for the airplane. Snow and ice adhering to the wings changes the shape of the wing cross-section, which raises the stall speed. Thus, because buffeting and stalling occur at higher speeds than they normally do, ice and snow on the wings may make it impossible for the plane to climb, creating an extremely dangerous situation.

There are no clear standards for how much accumulated snow on the wings will cause buffeting or induce stalling. In other words, there are no clear standards for the amount of accumulated snow on the wings that warrants postponement of take-off.

For that reason, there are accidents in which it was known that there was snow on the wings, but where the easy-going decision was made that it was not enough to hinder take-off; then, at the time of take-off, the aerodynamic lift required for lift-off could not be obtained, resulting in difficulty in operating the aircraft.

On 12 February 1979, the Allegheny Airlines flight 561 Nord 262 crashed 14 s after take-off from Benedum Airport near Clarksburg in West Virginia, killing two of the 25 persons on board and injuring eight. According to the National Transportation Safety Board, the cause of that accident was the loss of lift and control that resulted from the captain deciding to take off with accumulated snow on the main wings and on the tail wings.[13]

On 13 January 1982, the Florida Airlines flight 90 Boeing 737 took off from Washington National Airport during a heavy snowfall, with both the captain and copilot aware of snow having accumulated on the wings during stand-by. Immediately after take-off, the plane crashed into the railing of a bridge over the Potomac river, tumbled end over end, and plunged into the river, killing a total of 78 persons in the plane and on the ground. Both the accident and the laudable rescue efforts that followed remain in the memory.[14] One of the major factors in this accident was the fact that no clear standard for judging the danger posed by snow and ice build-up on the wings was provided by the manuals or other sources.

The following examples are similar to the two just described.

The weather was fine during the predeparture inspection of the exterior of the aircraft, but suddenly changed to a snowstorm, and the air temperature dropped to 28.4° F. The captain, feeling a tinge of concern after an approximately 4 min taxi to the end of the runway, confirmed from a window of the passenger cabin that there was just under 0.05 in of snow on the wings (appropriate information received). Nevertheless, he made the easy-going decision to take off, thinking that it was probably all right since there was no standard for snow and ice on the wings at time of take-off. The ice on the wing created difficulty in operating the aircraft, and the degree of danger in this incident was 'serious'.

In that incident, the pilot took prudent action. The pilot executed the take-off operation more slowly than usual, beginning to pull up at 115 knots (about 126 mile/h) and confirming that the speed was 80 knots (about 84 mile/h) at about 15 s after the beginning of the take-off run, the same as the usual speed. The pilot then continued the take-off under those conditions, but, when he began to pull up at 115 knots, the rudder felt heavy, the nose of the plane tended to drop, and 12 elevator trim units (devices for balancing the elevators of the plane) had to be used to balance the plane. This was due to the centre of gravity in this case; normally two or three units are used.

The reporter of this incident wrote that, although they did not climb sharply, but climbed with acceleration, they 'felt buffeting during the climb until the speed reached 130 knots (about 144 mile/h)'. The reporter also noted:

During the climb, visual observation of the wing tips indicated that there was no blow-off of moist snow by the wind pressure, but it remained in the form of icing, which probably changed the shape of the wings. We flew above the clouds until we reached the destination. The landing was normal, but a small amount of the ice remained after disembarkation.

In addition to snow, the background factors included a deficiency in the manuals, deficiency in the snow removal system, and the fact that there was no observation and reporting system in spite of the severe changes in weather. In Japan, there are no facilities such as the large snow and ice removal gates of the type used in Europe, which spray an ice removal solution over the entire body of the aircraft. Snow and ice removal is done mainly manually, greatly delaying departures.

Other cases include one in which strong buffeting was felt during the take-off climb and an abnormally large balance correction was necessary in the control (aileron) operation, and another in which the aerodynamic lift required for lift-off could not be obtained during take-off in a violent snow storm, and the climb rate was low and increases in altitude and speed were both extremely poor.

There are also cases such as the following. Although ice removal was performed before starting the engines, hail was falling during the approximately 2 min of taxiing to the take-off position. When inspecting the wings prior to take-off, a rather metallic texture could be seen, but take-off was begun because, if de-icing was performed again, it would not be possible to land at the destination airport within the curfew time. However, the rudder had a very heavy feel, and, with little levitation, as in the previous example, buffeting was experienced a number of times.

For these three example incidents, the reason for not reaching the stage of becoming an accident was 'good fortune' and the degree of danger was 'serious'. One reporter added in a supplementary note that 'I hope fully equipped snow removal facilities will be put into service soon'. Today, removal of snow and ice are performed relatively often, but reduced aerodynamic lift due to changes in wing shape caused by the viscosity of the de-icing solution is a new problem.

2.13 Danger of wind at take-off and landing

Crosswinds, tail winds, headwinds, and wind shear (sudden, violent changes in wind speed and direction) have strong effects on the operation of aircraft, particularly during take-off and landing.

Concerning crosswinds, to begin with, there is an incident in which, after landing, a strong crosswind caused the nose of an aircraft to veer sharply to the left and nearly off the runway. The captain used the steering mechanism of the plane to try to bring the aircraft nose back into alignment with the runway centre-line, but it did not have the intended effect and made the operation of the plane difficult. The copilot also hastily applied the right-side brakes, preventing the aircraft from leaving the runway. The degree of danger in this incident was 'serious'.

Another incident, which involved a 'considerable' degree of danger, occurred under near-minimum weather conditions, with a strong crosswind in addition to low visibility due to rain and fog and low clouds. The aircraft was making a ground-control approach to the right-side runway of two parallel runways.

The captain of the aircraft mistook the runway on the left to be the runway on which they were to land, and so was convinced that the copilot, who was flying the plane, was making the approach to the wrong runway. When the captain took over the control wheel to land the plane on the left runway, the copilot visually confirmed the right runway at an altitude of 500 ft. The copilot advised the captain to request clearance for the left runway which the captain was heading for. However, the captain ignored him and steered the plane in a reckless and forceful manner in an attempt to land on the runway on the right side. He made a 35° banked turn that resulted in a hard landing on the centre-line near the end of the runway.

In addition to the crosswind and minimum weather conditions,

background factors in this incident were the parallel runway config-
uration, the pilot training or check flights, excessive leadership by the
captain, and the captain's lack of confidence in the copilot.

In a 20 knot crosswind, the aircraft is being pushed sideways (which is
to say to the left in this incident, as the wind is coming from the left) by
33 ft every second, so the nose of the aircraft is pointing upwind until
the aircraft touches down so as to compensate for the crosswind during
the approach. Because the plane's attitude at that time brings to mind
the sidewise crawling of a crab, this manoeuvre is called 'crabbing'. At
the time of touchdown, the nose of the plane must be brought into
alignment with the runway centre-line. In this manoeuvre, which is
referred to as 'decrabbing', the windward wing of the aircraft is lowered
slightly, the fuselage of the aircraft is inclined, and the aircraft moves
downwind with the crosswind. Performing this manoeuvre in a variable
wind is technically quite difficult.

Furthermore, parallel runways do not always become visible at the
same time in an approach through fog or clouds. In this incident, the
left-side runway was the first to come into view as seen from the
captain's seat on the left side of the plane, because the nose of the plane
was pointed towards the left to correct for the wind. In the case of
parallel runways, because pilots have a strong instinctive inclination to
land at the first runway that comes into view, it is necessary to confirm
the runway by using electronic aviation safety facilities such as an ILS
or a longwave radio Automatic Direction Finder (ADF).

The following incident is an example in which a tail wind is a
background factor that leads to difficulty in aircraft operation.

Although there had been no wind prior to the start of the take-off
run, a 25 knot (approximately 40 ft/s) crosswind and a 10 knot
(approximately 16 ft/s) tail wind suddenly developed while the plane
was moving on the taxiway. The operation of the plane became so
difficult during the take-off run that it was not possible to track the
centre-line even using full rudder (the left–right direction control). The
degree of danger in that case was 'considerable'.

With a tail wind during approach to the runway, the ground speed of
the aircraft is equal to the airspeed plus the wind speed, so it is necessary
to increase the descent rate in order to maintain the appropriate
approach angle. As a result, the speed of the aircraft easily increases,
and, if the runway braking effectiveness is also lowered by rain or snow,
there is a possibility of overshooting the runway because a further
increase in the wind speed increases the ground speed, and thus extends
the landing distance.

What becomes a problem is that the ILS equipment is installed only on one direction of the runway. Considering only the wind direction, the landing should be made using the VOR alone from the opposite direction where the ILS is not installed, as landings are generally done flying into the wind (the VHF omnidirectional radio range displays the compass direction from the radio station using very-high-frequency radio waves). Because of the contending factors of heavy rain and poor visibility, however, the approach could only be made from the direction where the ILS equipment was installed, which had lower minimum weather conditions.

In such cases, waiting for the weather to improve or diversion to an alternative airport might also be considered, but usually landing is permitted if the tail wind is 10 knots (about 16 ft/s) or less, so the approach is made with the easy-going decision that it is all right.

There is also an incident whose degree of danger was 'serious' in which, although the runway could be discerned at the decision altitude, the altitude was too high to pull up the nose of the aircraft, so the plane touched down far beyond the normal touchdown point. What is more, there was a reduced braking effect and difficulty in manoeuvring the plane owing to hydroplaning (in which a layer of water forms between the tyres and the runway surface, greatly reducing friction). As a result, the plane nearly overshot the runway.

The aircraft in this incident received a communication over the company radio that indicated moderate rain, but, at the time of landing, it was raining hard enough to produce ankle-deep puddles. The background factors here were the difference between the weather information and the actual weather conditions, the lack of ILS, and the captain's route check flight (according to an examination conducted by the Ministry of Land, Infrastructure and Transport[15]).

Wind shear, a phenomenon that occurs locally in time or space and in which the wind direction or wind speed changes suddenly and violently, makes it difficult to control the aircraft because the effects of the wind increase and change in complex ways, resulting in extremely high danger. When wind shear is encountered at low altitude during take-off or landing, the pilot generally increases thrust to gain altitude while maintaining speed, but no operational procedure has been established for breaking away from wind shear to safety.

In the United States on 24 June 1975 there was an accident in which an Eastern Airlines flight 66 Boeing 727 encountered wind shear while landing at New York Kennedy Airport, resulting in loss of control. The plane struck approach lights and crashed, killing 115 people.[16] Nine persons received serious and light injuries. In addition to this, there were 27 other accidents involving wind shear in the 18 years from 1964 to 1982.[17]

The first accident in Japan for which wind shear was indicated as a contributing factor occurred at Naha Airport on 19 April 1984, when a Japan Asia Airways flight EG 292 DC 8 hit a landing approach light.[18] Several years before that, another plane encountered difficulty in manoeuvring in an incident of 'serious' degree of danger at the same runway of the same airport.

Other than this, the 430 incidents reported to the Japan Research Institute of Air Law by pilots include ten incidents in which wind shear was encountered, resulting in difficulty in decision and aircraft operation and a 'serious' degree of danger. In nine of those ten incidents, there was no advance information from the air traffic controller, the dispatcher (operations manager), or other such source concerning the occurrence of wind shear.

In the tenth incident, there was a report from the ground on a 5 knot east by northeasterly wind, but inference from the wind correction angle indicated a 40 knot westerly wind at an altitude of 200–150 ft. From that information, the presence of wind shear could be expected. Also, because there was information from the pilot of the plane that landed ahead of this flight, this is an example in which the pilot was able to cope with the wind shear even though it was difficult to control the aircraft.

Other meteorological conditions that have been background factors in operation errors are: passing through adverse weather conditions and front lines, fog and haze, clouds, cumulonimbus and hail, thunderstorms, and abnormal atmospheric pressure. Even with the highly developed aviation technology of today, we know well that weather conditions are deeply involved in the safety of air transportation.

2.14 Pitfalls of good weather conditions

In a discussion of weather conditions, good weather also demands special attention. Looking only from the perspective of manoeuvring the aircraft, good weather is the desired weather condition. When seen from the perspective of human factors, however, good weather conditions do not necessarily have a positive effect.

It is human nature to be drawn to things that are interesting or beautiful. One example of that is an incident in which, under conditions of extraordinarily good visibility on a clear and fine day, all three cockpit crew members were enjoying the sight of the beautiful scenery during a ground-controlled approach and made the descent above the assigned altitude. The error was noticed and corrected. Viewed in terms of the job of operating the aircraft, this was a lapse of memory and the good weather was a background factor to an error of forgetting. The degree of danger in this case was 'virtually none'.

This is one of the most valuable reports, as it shows in a straightforward manner an error that involves human factors. If this kind of incident is blamed on laxity, incidents that are based on factors that are intrinsic human characteristics to which everyone is easily susceptible will not be reported, and such factors will remain as hidden paths that lead to accidents. We use this information here with respect for the bravery of the persons reporting it. We ask the reader, too, to fully understand this purpose.

In another case, a plane was making a night-time approach to Haneda International Airport in Tokyo with visibility so good that the shape of the land around the airport could be distinguished from the vicinity of Oshima Island. It was difficult to see the runway lights amid the flickering of the street lights around the airport, making it difficult to distinguish the runway until the copilot pointed it out in the air over Shinagawa, resulting in delayed action and a 'fairly little' degree of danger. The reason is that the runway lights can be seen easily from the front, but are not easily distinguished from the side or at an angle.

2.15 Latent danger in ever-advancing aviation safety facilities and instruments

The most commonly known aviation facilities are airport and aviation safety facilities. Aviation safety facilities are designed to aid the pilot in flying the aircraft by means of radio signals, lights, colours, and shapes.

Radio navigation aid facilities (aviation safety facilities) include the following:

- ILS (instrument landing system);
- VOR (very-high-frequency omnidirectional radio range, a radio beacon that indicates the compass direction from a radio station using very-high-frequency radio waves);
- NDB (non-directional beacon, a radio beacon that indicates the compass direction from the radio beacon using low-frequency radio waves);
- the on-board ADF receiver (a low-frequency radio wave automatic

direction finder, an instrument that indicates the direction from which a radio signal arrives on a clock-like display).

The ILS displays a continuous visual representation of the position of the aircraft fuselage with respect to the runway centre-line and descent angle on an instrument in front of the pilot. It can be used in conjunction with the automatic controls of the aircraft. Even in the case of manual control, rapid correction of deviations in course and approach angle is possible. Currently, the ILS has been installed as the main approach and landing equipment in various countries. In Japan, however, it has not been installed in all runways of class 1 airports (international airports), and virtually no class 3 airports (local airports)[19] have these facilities.

With ground-controlled approach, in contrast to ILS, an air traffic controller guides the aircraft while monitoring a ground radar screen up to the time just before landing. The air traffic controller gives instructions to the pilot by radio while watching the trace of the aircraft on the radar scope. Upon receiving the instructions, the pilot controls the aircraft accordingly. Thus, the control operations are slower than in the case of ILS. Accordingly, ILS facilities are necessary for maintaining a safe approach by immediately coping with sudden deviations from the course or descent angle as a result of wind shear or other such phenomena.

The incident described earlier, involving a landing on runway 18 of Naha Airport, was similar to an accident on 19 April 1984 in which the aircraft struck an approach light after encountering wind shear. In both cases, the aircraft suddenly encountered wind shear in a violent rainstorm when landing at the same runway of the same airport without any information or advice. Control of the aircraft was lost momentarily, and, although one tyre burst when the plane touched down on the landing gear of one side, it lifted off again successfully, with a degree of danger of 'serious'. The background factors were rain, wind shear, the lack of advance information, and the absence of an ILS system.

As for runway 18 of Naha Airport, although it was believed to be technically difficult to install ILS facilities because of the Futemba military base, now, as the base problem is being resolved, it is expected that ILS facilities will be installed.

Because the ILS displays the approach angle accurately, it is possible for the aircraft to touch down accurately at the touchdown point. This not only makes it possible to prevent overshooting short runways, it also makes it possible to avoid mistakenly landing at the wrong runway when approaching one of two parallel runways and only one runway is visible because of low visibility. Incorrect actions regarding whether or not the visible runway is the one on which the plane is supposed to land can be avoided through confirmation based on the ILS frequency or the identifying Morse code.

At Los Angeles International Airport there are two pairs of parallel runways in the same direction, for a total of four parallel runways. Those runways are named 24L (left), 24R (right), 25L (left), and 25R (right). Because the names overlap and have similar pronunciations, it is easy for the names to be heard incorrectly. Particularly under low-visibility conditions when none of these runways can be seen, the pilot, after a long flight, has a strong inclination to take the first runway that comes into view to be the one on which to land.

There is an incident in which a plane that was cleared for an ILS landing on runway 24L was making its approach. When the pilot was attempting to line up the aircraft with the centre-line of the runway that he had seen (which was actually runway 25R), he was able to confirm that the runway was 25R rather than 24L when he saw that another aircraft that had been cleared to land at runway 25R was approaching. The incident thus ended with a 'considerable' degree of danger. In such cases, reconfirmation by means of the ILS frequency or identification code can prevent this kind of serious error.

With VOR or NDB, it is not possible to know the glide path angle from an instrument on the aircraft. Accordingly, in the case of airports or runways that do not have ILS facilities, the pilot must rely on his visual judgement of the glide path angle. Because that is less accurate than when ILS is used, the burden on the pilot is greater. Under weather minimum, a VOR or NDB approach must become higher than an ILS approach. Particularly when visibility is low owing to low-hanging clouds or fog, it sometimes becomes impossible to land, thus becoming a cause of lower service efficiency than in the case of airports using ILS.

2.16 Airport markings and landmarks as important factors

Airports consist of runways, taxiways, parking aprons, and so on. Just like the markings on highways, these airport areas have signs and markings such as stop lines and taxiways, and centre-lines. These markings and signs are obviously important to the safe operation of aircraft. Accordingly, any problem with these signs or markings will affect the decisions and actions of the pilot. While there is a tendency to take danger on the ground less seriously than danger in the air, being on the ground is no reason for inattentiveness, as we learned from the accident involving an on-the-ground collision at the Tenerife Airport on 27 March 1977. When the aircraft is on the ground, markers also serve an important purpose.

Each runway has multiple taxiways, and, in the case of parallel runways, the maze of taxiways is even larger. Moreover, the shapes are similar, so, if the markings are not clear, difficulty in deciding which taxiway the plane is on or where to turn is not unusual.

On 7 December 1983, an Avianca Airlines domestic flight 134 DC-9 collided on the runway with an Iberia Airlines flight 350 Boeing 727 that was taking off at the Barajas Airport in Madrid, Spain. Both planes burst into flames in this accident, killing 93 persons. This accident happened because the pilot of the Avianca plane mistook the taxiway owing to improper and insufficient marking and taxied across the runway.[20]

Before this accident, pilots had been petitioning the Barajas Airport authority to install stop line markings and lights at the intersections of taxiways and runways. Although the current situation is unclear, one and a half years after this accident occurred there was said to be no change in the circumstances and no budget established for that purpose.

Another accident occurred on 3 December 1990 at Detroit Airport, which has been called the airport that 'makes one's hair stand on end' because of its maze of taxiways between runways that intersect at angles. In dense fog, a Northwest Airlines flight 1482 DC-9 strayed onto a runway that was in use and collided with a Northwest Airlines flight 299 Boeing 727 making its take-off run. Eight people died in this accident. The US National Transportation Safety Board pointed to improper marking as a contributing factor in this accident also.[21]

Before this accident, a similar incident had occurred in Japan. Just before reaching the stage of becoming an accident, the landing aircraft escaped a dangerous collision by a hair's breadth by executing a go-around. The degree of danger in this incident was 'serious'. The reason for recovery was action by the other aircraft. Distracted by conversation with other crew members, the pilot mistook the runway for the taxiway, both of which were under construction, and, making an easy-going decision, mistakenly moved out into a runway that was in use (an active runway.)

Signs or markings on the ground surface, such as arrows or X marks that indicate that taxiways or runways are closed, can become difficult to see, not only because of hindrances to visibility such as heavy rainfall, falling snow, or ground fog but also because of night-time darkness or accumulated snow.

Lights as markers are not effective in the daytime; painted markings are easier to see. At night, however, the reverse is true and painted markings are extremely difficult to see; if there is accumulated snow, painted markings sometimes cannot be seen. In addition, light reflection from wet runways can make even newly painted markings hard to see, and old markings that are supposed to have been covered up by new markings sometimes become visible.

Pilots are normally informed of changes in the guidance lines (arrows) through NOTAM (the notice to airmen) or other such means, but repair work on taxiways is done sequentially, one section at a time, so the markings are repainted in accordance with the progress of the repair work. Although the old markings are obliterated by overpainting with black paint, reflections from the taxiway lights at night can make the old markings appear like the markings that are now in effect, and there are incidents in which aircraft followed old markings, resulting in close spacing of aircraft on the parking apron and a 'fairly little' degree of danger. Wrong conviction caused by seeing incorrectly leads to incorrect action.

There is also an incident in which the pilot was following the wrong markings 'because the old guidance lines had not been completely obliterated, but were easily visible', but noticed the mistake when the captain pointed it out. The degree of danger in this case was 'virtually none'.

In another incident of 'fairly little' degree of danger, a plane that was moving on a taxiway for take-off at night was instructed to stop before entering the runway. The crew was informed that the taxiway stand-by position marker was difficult to see and so was devoting much attention to spotting the marker. In spite of that, the stop line could not be found from the captain's side; the copilot, seeing the line vaguely, hit the brakes and stopped the plane. In this case, the flow of the information process was delayed decision followed by excessive action due to not seeing or difficulty in seeing. The reason for recovery was the advice and action of the copilot.

The pilot who reported this incident expressed the desire for the use of highly reflective fluorescent paint or the installation of lights so that the markings can be seen easily at night. Clear markings at the intersections of taxiways and runways, in particular, are important in the prevention of accidents, and are strongly demanded by pilots, just like at the Barajas Airport.

Highly accurate navigation is possible with INS (inertial navigation system) equipment, but for INS to function properly, it is necessary correctly to input the latitude and longitude of the parking spot while the plane is parked there. If the wrong latitude and longitude values are input, even if the values are for a nearby parking spot in the same airport, there will be a slight error in the present location of the plane. If the plane is flown with the wrong position having been input, the error will increase with the distance the plane flies. If the pilot notices the error during position notification, then the error can be corrected.

There is also an incident in which the wrong latitude and longitude of a similarly numbered parking spot at a different airport were entered when confirming the current location because a completely unfamiliar error message number appeared on the INS operation display panel after the current location data were input. In this incident, a seeing incorrectly led to a wrong conviction and an incorrect action that resulted in a 'virtually none' degree of danger. The background factors included fatigue due to a short one-night on-the-ground rest in 3 days of flight that involved many landings.

Now, in Japan, the input of latitude and longitude is performed by looking at a binder-type table that is distributed to the crews. Because of this, there is a danger of making an input error by incorrectly seeing the parking apron number by turning to the wrong page or because the wrong page has been inserted.

As one measure to prevent these kinds of mistake, displaying the location of the parking spot by latitude and longitude in a manner easily seen by the pilot from the aircraft instead of displaying only the parking spot number is increasingly used by airports in countries other than Japan. Furthermore, some airports are adopting an optical guidance system to prevent aircraft from proceeding to the wrong parking spot after landing.

2.17 Importance of weather information

Even though aircraft equipment and airport facilities are being improved, that alone is not sufficient to overcome the problems of safety. Weather conditions – and particularly weather conditions that change from minute to minute in the vicinity of airports – must be reported to pilots in real time, because they greatly affect the operation of aircraft.

The World Meteorological Organization (WMO) makes special observations of wind direction and velocity, visibility, runway visual range, cloud coverage, thunder and lightning, hail, and other conditions, and determines which weather conditions should be reported to aviation organizations. In actuality, however, there is an insufficient number of personnel stationed to make the weather observations, and no equipment for observing wind shear[22] and other such phenomena are installed, so observations are not made. This is an important background factor in the occurrence of highly dangerous incidents and accidents.

In the Civil Aeronautics Law of Japan, the Minister of Land, Infrastructure and Transport stipulates that weather information must be provided to pilots. In addition to obtaining weather information directly from the air traffic control of the airport at which the plane is landing or taking off, pilots can also obtain it from three sources: ATIS (air traffic information service, an airport information service that continuously broadcasts information on changes in clouds, wind, air currents, runway use, etc.), AEIS (aeronautical en route information service, a communication service for aircraft en route and weather information broadcasting service), and VOLMET (volume meteorological, periodic weather information broadcast by shortwave radio). When this information is insufficient, it can be supplemented by information from the dispatchers of airlines or the branch offices of the pilot's own company.

Concerning the ATIS broadcasts, it has been pointed out by some pilots of other countries that it is difficult to understand the broadcasts. There are also numerous incidents in which the fact that such information was insufficient, inappropriate, or differed from the actual situation at the time was involved as a background factor.

Next, let us look at an example in which a difference between the reported weather and the actual conditions resulted in an incident whose degree of danger was 'serious' and whose reason for recovery was 'good fortune'. During a night-time landing, the visibility suddenly deteriorated owing to a rainfall that was so heavy that the rain removal system – equipment that uses compressed air to blow raindrops off the windows – became ineffective. The pilot was thus unable visually to confirm the runway conditions that would affect the coefficient of friction of the aircraft tyres at the time of landing, and had only the weather information reported from the ground to rely on in making the landing.

The touchdown was normal, but the runway did not have centre lights, so, even though the lights on both sides of the runway could be seen, it was not possible to know whether or not the plane was moving in the centre of the runway. The plane was sliding and yawing (lateral movement of the aircraft nose) as though hydroplaning, and in the next instant the nose of the aircraft made a large swing to the left, so the pilot used right full rudder to stop that deviation.

Although this incident fortunately did not develop into an accident, it brings to mind a related case in which an Eastern Airlines flight 576 Boeing 727 slammed into the ground during an ILS approach to Raleigh/Durham Airport in North Carolina on 12 November 1975. The fuselage of the plane was seriously damaged and eight of the 139 people on board suffered serious and light injuries.

Concerning this accident, the accident investigation report of the US National Transportation Safety Board stated that the weather was worse than the forecast, and, as the plane passed the decision altitude, the rain became stronger and 'the windshield wipers, the most classical mechanism on jet aircraft, were barely operating in the midst of the water'.[23]

In both this and the previous example, the difference between the weather information reported to the pilot and the actual weather conditions was a background factor. In either case, the actual conditions warranted a missed approach procedure, but the pilot, depending only on the weather information reported from the ground, was delayed in deciding to do a missed approach, resulting in reckless operation.

The pilot who reported this incident wrote that 'When we entered the parking apron, the rain stopped and the situation of 3 min earlier seemed unbelievable, but, for an instant, a chill ran through my body'.

In an incident described earlier in this book, an aircraft that was approaching Naha Airport become difficult to control after encountering wind shear in heavy rain, and burst a tyre when it touched down. From this example we can see that, if an aircraft encounters severe and sudden changes in weather when approaching a cold front or thunderclouds without prior information, the result can be difficulty in controlling the aircraft and an extremely dangerous situation.

2.18 Close relationship between weather forecasts and fuel on board

Another case in which a difference between the weather forecast and the weather conditions actually experienced greatly affects aviation safety is the issue of fuel on board, or the amount of fuel the plane is carrying.

In one such incident, fine weather had been forecast for the destination at the time of landing, so the plane departed with relatively little fuel for flying a holding pattern while waiting to land. However, the weather at the destination airport suddenly changed and the plane was not able to land. The pilot began a holding pattern after receiving a communication from the ground that landing would be 'OK in 30 min or so'. After 30 min had passed, however, there was no improvement and the plane was unable to land even though two or three approaches were attempted.

It was decided to divert to the alternative airport, but that airport was closed because of an accident, so there was no choice but to try to land at the destination airport. Fortunately, the plane was able to land, but, if it had not been able to land and a missed approach had been necessary, there would not have been enough fuel to do so.

The degree of danger for this incident was 'serious'. Although the incident involved a chain of three events, the information source was the critical phase in the first and second events in that chain. Moreover, the difficulty in diverting to another airport in the action phase of the third event was the incident critical phase that determined this incident. The background factors were the sudden occurrence of the closure of the alternative airport, the low remaining fuel on board, the flaw in the information source in the first event (that is, the weather information for the destination airport that was received at the time of departure), and the flaw in the weather information prediction of the second-event information source that the weather conditions would be 'OK in 30 min or so'.

This incident did not reach the stage of becoming an accident because the plane was able to land on the last approach attempt, but, if it had been necessary to execute a missed approach at that time, there would have been danger of a forced landing or of the aircraft crashing, because 'there was nearly zero fuel left'. The reason for recovery was 'good fortune'.

An accident that is highly similar to the above incident occurred on 25 January 1990, when an Avianca Airlines (Columbia) flight 52 Boeing 707 that had departed from Bogota bound for New York Kennedy International Airport via Medillin crashed into a hill 15 miles north of Kennedy International Airport. Of the 158 persons on board, 73 were killed and all of the survivors suffered serious injuries.[24]

After holding for 1 h and 17 min owing to deteriorating weather, clearance was given to land, but, when executing a missed approach because of encountering a strong headwind and wind shear, there was not enough fuel left to hold around the airport or to execute another landing, let alone divert to the alternative airport. That is to say, the immediate cause of this accident, the stopping of the engines of the plane, occurred because the fuel was used up.

In the ground briefing held before the departure of the aircraft, the dispatcher did not explain to the pilot the most recent weather conditions at New York, nor did he mention anything about the possibility of delays due to adverse weather conditions or diversion to the alternative airport. Moreover, there were no arrangements concerning the fuel required for a diversion.

The captain of the aircraft instructed the copilot to send an 'emergency' message so that the air traffic controller would take measures for a priority landing, but, not recognizing the point, the copilot nevertheless simply sent the message '... we are running out of fuel'. Furthermore, when a short while later the air traffic controller gave instructions for the plane to climb, the copilot responded simply with 'No good. We are out of fuel'. That failed to convey the seriousness of the situation, and the air traffic controller did not consider the aircraft to be in an emergency situation.

The accident report of the US National Transportation Safety Board (NTSB) stated that, concerning the management of the situation by the air traffic control of that airport, which subjected the aircraft to an abnormally delayed landing, the responsibility lies with the Federal Aviation Agency. It found no problem with the measures taken by the air traffic controller with respect to this aircraft.

One member of the NTSB, Mr Jim Burnette, and Mr M. H. E. Real, the Colombian Civil Aviation Safety Administration Officer, pointed out that the fact that the latest wind shear information was not provided to the pilot of the Colombian plane was a problem on the part of air traffic control and contributed to the accident.

Whatever the case, as far as the weather conditions at Kennedy International Airport are concerned, the pilot of that plane was not informed of the latest information either by the dispatcher before departure or by air traffic control as the plane approached the airport.

2.19 Point where mechanical problems develop into an accident

The aircraft is handed over to the pilot after the pilot confirms that the maintenance has been done properly before departure. Sometimes, however, problems are not discovered at the time of this predeparture check.

If one engine is not operating or one instrument has failed, the problem does not usually develop into an accident. However, if the pilot's attention is focused on that problem, it detracts from his ability to attend to other matters and thus creates a danger of an accident developing. It is human nature to focus one's attention on something that is interesting or something that is of concern ('excessive focus'). A typical example of that is an accident that occurred on 29 December 1972, when an Eastern Airlines flight 401 Lockheed 1011 was landing at Miami Airport.[25]

The problem was simply that the bulb of the green lamp of the nose gear indicator system (an indicator lamp showing that the front landing gear has been completely lowered) was burned out, but, when the pilot turned around to ask the engineer about the status of the nose gear, his body brushed against the control column, unintentionally deactivating the automatic control system. However, failing to notice what had happened, the two crew members paid no attention at all to the flying of their own aircraft and also did not monitor the instruments. The aircraft continued to descend until it struck the ground, killing 103 persons (73 persons were rescued).

An incident similar to that accident occurred when the captain of a plane set the selector switch so that the data from the flight data computer on the copilot's side would be displayed on the instruments on the captain's side during the take-off, because the flight data computer on the captain's side was not functioning. By mistake, he turned off the switch on the copilot's side, and so, for a moment, neither of the two computers of the plane was operating. Because the two pilots were concentrating on this problem, the decision to retract the landing gear and the operation of retracting the gear were forgotten. The pilots became aware of that error because of the slow acceleration and vibration of the plane. The degree of danger for this incident was 'fairly little'.

In another similar incident that occurred during a training exercise, the instructor captain and the copilot got the impression that the Automatic Direction Finder (ADF) reading was wrong during a climb. Because of their fixation on that, they forgot to retract the flaps and failed to notice it, even though they were exceeding the speed limit. The chain of events was broken when the flight engineer pointed out the situation, so the degree of danger was 'virtually none'.

There is also an incident in which the pilot and copilot were concentrating on a problem with the flaps during an approach, and, because they were fixated on that problem, they overshot the ILS course and, even though they hurriedly made a correction to bring the nose of the aircraft back onto the course, they still overshot, resulting in a degree of danger of 'fairly little'. A background factor here was the unclear division of duties among the crew members because of insufficient leadership by the captain.

In another reported case, the attention of the entire crew was absorbed by a trivial problem with the Profiler Meteorological System (PMS, a computerized navigation control device) during an approach because of unfamiliarity and insufficient experience with that system. As a result, they forgot to set the barometric altimeter setting value (QNH, altimeter subscale setting to obtain altitude above sea level). They noticed it when they received a request for reconfirmation of the QNH setting from the air traffic controller. The degree of danger was 'fairly little'.

There is also an incident of 'fairly little' degree of danger in which decision and action were delayed by attention being focused on a compass abnormality while the aircraft was moving on the ground. Just as the plane was about to cross over the stop line and proceed across the runway, the captain noticed the situation and quickly applied the brakes.

In yet another example, the captain and flight engineer were fixated on an abnormality in the landing gear during take-off and so were late in beginning the initial turn of the Standard Instrument Departure (SID, multiple typical departure routes for the instrument flight set for each runway of each airport, including points over which the aircraft should pass, altitudes, etc.). In this case, the copilot was deliberately monitoring the situation outside the plane and the instruments, so it was possible to continue with a normal flight after that. Because of a good division of duties, the degree of danger for this incident was 'virtually none'.

When mechanical problems lead to errors, it is most often a matter of excessive focus on the mechanical problem; this leads to fixation on a decision or action, which results in the forgetting or delay of attention to another matter. In adverse weather conditions, however, it is highly possible for the failure of navigation instruments, such as abnormality in the ILS system or an inoperative or improperly operated altimeter or speed indicator, to lead to an accident. However, if crews are given training in dealing with failures of aircraft systems relatively often, the division of duties is clear, and the communication among crew members is good, then the degree of danger is greatly reduced. In these kinds of incident, the degree of danger differs according to the point in time at which the state of excessive focus ends.

2.20 Pilot training or check flights that increase stress and workload

The examinations and training exercises that are conducted for the pilots of commercial airliners include route checks, training flights to obtain qualification for different types of aircraft, periodic check flights (every 6 months for captains and every year for copilots) stipulated by the Civil Aeronautics Law of Japan, and the training and examinations for job promotions and periodic training conducted by airline companies.

For these, except for some of the periodic examinations and route checks, an examiner or inspector of the Civil Aviation Bureau of the Ministry of Land, Infrastructure and Transport is present at the final stage of the examination, as is the inspecting pilot of the airline company. That is generally called the inspection flight. Although, recently, simulators have been used for training and inspection, final route check flights are conducted in actual aircraft and furthermore on scheduled flights with passengers on board.

Local flight training (mainly approaches, take-off and landing operations, emergency countermeasures, etc., which are conducted in the vicinity of an airport) includes training in emergency operations, such as when an engine stops or a hydraulic system becomes inoperative. These exercises thus involve not only a heavy workload but also a high degree of latent danger.

While not involving in-service aircraft, there have been accidents during training, such as on 27 February 1965, when a Japan Airlines CV-880 trainer crashed and burned just short of the Iki runway, injuring one person;[26] in August 1966, when a Japan Airlines CV-880 trainer ran off runway 33 of Tokyo International Airport during a take-off run and caught fire, killing the pilot and four others;[27] and on 24 June 1969, when a Japan Airlines CV-880 trainer crashed after take-off during training at Moses Lake Airport, killing three of the five persons on board.[28]

There have also been accidents including the airlines of countries other than Japan, such as when a Delta Airlines CV-880 trainer crashed during a check flight on 23 May 1960, killing all five persons on board. Many other accidents have occurred during training in the same type of aircraft.

An example incident of this type occurred during a training exercise for landing with one engine out. Engine Nos 2 to 4 were all throttled down to minimum thrust and then the thrust was reversed. The thrust lever for the No. 1 engine (the lever that controls the engine output power, similar to the accelerator pedal of an automobile), however, was still in a slightly forward position, thus creating a slight forward thrust. With the thrust of the other three engines reversed, this resulted in a large difference in thrust between the left and right side of the aircraft which caused the nose to swing far to the right. The plane thus came close to the right edge of the runway, resulting in a 'considerable' degree of danger.

In addition to pilot training or check flights, the background factors in this incident were above-normal workload because of the engine-out training, the fact that both the captain and the pilot trainee were concentrating on the situation outside the aircraft, and the fact that they did not see the position of the thrust lever. On the advice of the flight engineer, the thrust lever for the No. 1 engine was moved to the minimum thrust position and the attitude of the plane was corrected.

Also, there is an example in which, in a situation of excessive tension during pilot training, the captain and copilot both forgot to lower the landing gear even though the aircraft had passed the decision altitude. The recovery was made when the flight engineer advised the pilots of the situation, and the degree of danger for this incident was 'considerable'. There is also a similar incident in which the operation of lowering the landing gear was forgotten and the degree of danger was 'serious'. The crew did not hear the warning buzzer, and, when the captain trainer noticed the situation and executed the go-around, the aircraft was a mere 8–12 in from the ground.

The background factors in incidents such as this most often include the excessive stress and tension on the pilot trainee and the heavy workload of the captain trainer.

Also, although communication with air traffic control is usually done by the copilot, during pilot training the captain trainer must handle the communication with air traffic control, keeping watch on the situation outside the plane and other such duties of the copilot in addition to his own duties as an instructor. There are occasional cases in which the trainer captain is concentrating his attention on instruction and, unable to divide his attention, forgets to obtain clearance from air traffic control and lands the plane without clearance. During pilot training, the instructing captain can easily become meddlesome in the aircraft operation, but meddlesomeness hinders the full development of the trainee's abilities.

2.21 Unexpected changes of instructions as a cause of mistakes

Article 95 of the Civil Aeronautics Law of Japan stipulates that the operation of an aircraft must follow the instructions of the Minister of Land, Infrastructure and Transport, or, in actual practice, the instructions of the air traffic controller (for details, see Chapter 3). However, the instructions of the air traffic controller may change during the approach to the airport, depending on relations to the operation of other aircraft, weather conditions, and so on.

These kinds of instruction are given for the safe operation of the aircraft (collision prevention and orderly flow), but, if the air traffic controller's instructions are changed midway, the pilot must hastily change the control and operation plan that he has constructed in his mind on the basis of the original instructions. If there is a change in the runway on which to land just before the landing owing to a change in the weather conditions, it also involves a change in the aviation safety facilities, the direction and altitude of the approach, the missed approach procedure, and so on. Therefore, it is necessary to change the control and operation plan.

When making an approach and landing, the workload of the cockpit crew is high, and, if there is a change in the runway on which to land, particularly in the case of a crew fatigued by a long flight involving jet lag, the original flight plan constructed in the pilot's mind remains latent there. Because of that, it can easily happen that the latent idea becomes a preconception and matures into a wrong conviction when making the approach and landing at the runway specified by the new instructions, resulting in a mistake in operation.

Particularly when the landing runway has been changed, in the case where it is necessary to land from the opposite side, it is difficult completely to switch one's left–right thinking. There is thus a danger of running off the runway because the opposite and wrong operation for compensating for a side wind is performed. When a side wind factor is in contention with fatigue factors and an unexpected change during approach after a long flight with jet lag, the captain's workload is extremely high, as he must give instructions to the other crew members concerning changes in the decision altitude, the frequency of the ILS, course settings, and so on, and the missed approach briefing must be redone.

There is an example of this kind of incident in which the degree of danger was 'considerable'. The weather information received when the descent for landing was begun indicated a 15 knot (about 25 ft/s) crosswind from the right at the ground surface. At an altitude of about 10 000 ft there was a change in runway, and, because the pilot mistakenly made the opposite correction for the crosswind at the time of landing, the plane nearly ran off the runway. The heading of the aircraft was kept to the left (upwind) until the aircraft touched down so as to compensate for drift resulting from crosswind during descent. At the time of touchdown, the nose of the plane must be brought into alignment with the runway centre-line, so the windward wing of the aircraft is lowered slightly, the fuselage is tilted, and the aircraft moves downwind with the crosswind. Although during approach the pilot continued his descent with the nose of the plane pointed to the left, which is the direction that the wind was coming from, there was still a residual awareness in the pilot's mind that the wind was coming from the right, based on the original approach plan before the change in runways. When the plane touched down, instead of lowering the left wing, as he should have, he mistakenly lowered the right wing, causing the aircraft to drift to the right.

In another reported case, during an approach to two parallel runways, the runway to be used was changed from the right to the left runway, and, although the plane should have turned right after landing to proceed to the parking spot, the pilot had it still in his mind to turn left in accordance with the briefing that was held prior to the change in instructions. Thus, in spite of the change in runways, the pilot did not switch the locations in his mind and the idea of turning left stuck in his mind. Fortunately, an instruction to turn right was given by the air traffic controller, so this incident ended without problem and a danger level of 'virtually none'.

Unexpected changes in instructions from the air traffic controller are not restricted to which runway is to be used; flight route and altitude are also subject to change.

In one incident, both pilots entered the way point of the original route into the INS, in spite of a change in the flight route from the original flight plan that was made while the flight was in progress. Recovery was made on advice from the flight engineer. In another incident, the assigned altitude that was received on the ground was changed during the climb, but both of the pilots attempted to climb to the original altitude. In each case, the flow of information processing involves a firmly rooted preconception that leads to a wrong conviction and an incorrect action with a danger degree of 'virtually none'.

There is also an incident in which an instruction for a turn that was different from the usual was received from air traffic control, and, because the pilot's attention was on that, he forgot to lower the landing gear. The oversight was not noticed until late, when the plane was at an altitude of between 200 and 300 ft. The background factors included heavy workload and the degree of danger was 'serious'.

In a similar example, when beginning a missed approach because of snow, the runway became visible so the pilot intended to land. However, the copilot did not extend the landing gear, having forgotten that the landing gear had been retracted for the missed approach and was still up. The fact that the gear was not down was not noticed until the plane was near an altitude of 100 ft and the degree of danger was 'serious'.

The background factors that have been found in the 430 incidents related to operating the aircraft include the already described personal factors of crew interpersonal relationships and meteorological conditions. Aside from these items, the following should be considered background factors (see Table 2):

- snow removal capability;
- airport lighting;
- inappropriate ATC instructions;
- other nearby aircraft;
- inappropriate or insufficient advance information, including NOTAM (flight information), the AIP (aeronautical information publication, information published by the government of each country), and charts;
- parallel runways;
- sudden occurrence;
- diversion (change in landing site);
- low remaining fuel;
- scheduling delay;
- time restrictions on take-offs and landings (curfews);
- heavy or light workload;
- long-distance flight, time-zone differences (jet lag), and many take-offs and landings per day;
- operation management policies (company policies: keep time schedule, economy of operations, and labour relations).

Other than these, factors such as low oxygen, back-light, similar call signs, radio frequency for communication with air traffic control, and birds were also found, but, as the number of cases were few, they were reluctantly omitted from the quantification method III analysis.

Table 2 Factors for quantification method III, pilot-related human-factor incidents

Total number of incidents: 430

Total number of events: 527

	Categorical No.	Category*	Frequency
Time of occurrence	1	Day	189
	2	Dusk	31
	3	Night / moonlight	80
	4	Dawn / early morning	33
Flight phase	5	On ground (before take-off)	43
	6	Take-off	48
	7	Climbing	62
	8	Cruise	41
	9	Descent	44
	10	Approach	192
	11	Landing	47
	12	On ground (after landing)	21
Meteorological factors	13	Fair / clear	45
	14	Crosswind / tail wind / wind shear	36
	15	Abnormal atmospheric pressure	5
	16	Cumulonimbus / thunderstorms / hail	10
	17	Fog / haze	11
	18	Rain (heavy / moderate / light / showers)	22
	19	Snow (heavy / moderate / light / showers)	29
	20	Icing (heavy / moderate / light)	5
	21	In clouds	14
	22	Adverse weather conditions (below minimum / front line passing)	7
	23	Weather minimum (low ceiling / low visibility)	10
Aircraft conditions	24	Malfunction / mechanical problems	23
Navigational aids (Navaids) and airport safety facilities	25	Weather observation and reporting system	14
	26	No ILS (insufficient navigation aid)	21
	27	Snow removal capability	13
	28	Runway condition (dry, wet, slippery, snow, ice, slush, etc.)	18
	29	Runway and taxiway marking and signs / airport lighting	13
Operation management policies (company policies)	30	Economy of operations / keep time schedule / labour relations	20
Special background factors	31	Unexpected change in ATC clearance (route, runway, altitude), sudden occurrence	30

		Categorical No.	Category*	Frequency
		32	Check / training flight	32
		33	Scheduling delay / curfew / diversion	15
		34	Heavy workload	35
		35	Light workload	10
		36	Long-distance flight / frequent take-off and landing, jet lag, low remaining fuel	31
		37	Inappropriate or insufficient advance information (WX, NOTAM, etc.)	41
		38	Other nearby aircraft	39
		39	Inappropriate ATC instruction or information	45
		40	Parallel runway	13
Crew co-ordination		41	Good condition	70
		42	Overreliance / interdependence / excessive familiarity among crew	34
		43	Meddlesomeness	12
		44	Excessive concern / misplaced deference	29
		45	Unreliability, graveness	17
Personal factors	Experience	46	Insufficient experience / unfamiliarity	29
	Physiological factors	47	Fatigue / lack of sleep	37
	Psychological factors	48	Inattentiveness (state of reaction after stress) / tedium	27
		49	Impatience / fluster	58
		50	Stress, tension	23
Phase I (information source)		51	Instruments	121
		52	Outside visual information	150
		53	Documents (charts, manual, NOTAM, AIP, checklists, weather maps, etc.)	27
		54	Weather information [ATIS (audio weather info.), company radio]	26
		55	Crew communications (captain's commands and directives, other crew voice)	32
		56	ATC instructions	98
Phase II (information receiving)		57	Appropriate (information receiving)	140
		58	Did not see / hear (no input)	27
		59	Failure to see / hear / notice	31
		60	Saw incorrectly / heard incorrectly / mistaken impression	83
		61	Vacillation (information receiving)	4
		62	Excessive focus	41
		63	Preconception	41
		64	Delay (information receiving)	4
		65	Could not see / hear, could not see clearly / hear clearly	41

	Categorical No.	Category*	Frequency
Phase III (decision)	66	Appropriate (decision)	61
	67	Forgot (no decision)	74
	68	Easy-going, assumption	60
	69	Wrong conviction / hasty conclusion / one's own interpretation	161
	70	Vacillation (decision)	17
	71	Fixation	26
	72	Unreasonableness / high-handedness	14
	73	Delay (decision) / insufficient consideration	25
Phase IV (action / instruction)	74	Appropriate (action / instruction)	36
	75	Forgot (no action / instruction)	97
	76	Incorrect action / instruction, inappropriate action / instruction	198
	77	Loss of stability	8
	78	Fixation, persistence action	4
	79	Recklessness, forcefulness, excessive operation	60
	80	Delay (action / instruction)	29
	81	Difficult action or instruction / inability to act or instruct	59
Incident critical phase	82	Information source	39
	83	Information receiving	106
	84	Decision	127
	85	Action / instruction	135
Reason for recovery	86	Instruments, wearing lights and horns	14
	87	Outside visual information	15
	88	ATC advice	33
	89	Captain's advice, takeover	42
	90	First officer's advice, action	44
	91	Second officer's, flight engineer's advice, action	25
	92	Recovery by operator	62
	93	Good fortune	47
	94	Averting operation of other aircraft	8
Degree of danger	95	Virtually none	88
	96	Fairly little	110
	97	Considerable	111
	98	Serious	67

* The categories referred to in quantification method III; that is, the factors in this book.

2.22 Accidents averted by good fortune

The reasons for recovery in the 430 incidents that were studied include action by noticing one's own mistake, advice or action by the copilot or another pilot in the aircraft, advice or action by the flight engineer, decision or action by the captain, instruction or advice by the air traffic controller, outside visual information, instrument confirmation, operation of a warning device, and the averting operation of other aircraft, but, in addition to these, there were incidents that did not develop into accident out of good fortune.

The term 'good fortune' refers to cases in which the incident did not develop into an accident in spite of the fact that averting action was not taken or averting action could not be taken. Under normal circumstances, such incidents would have been expected to reach the stage of becoming an accident, and their failure to evolve into accidents was a matter of luck.

Examples include the case of the accident described earlier in which an All-Nippon Airways jumbo jet was brushed by a Self-Defence Forces MU-2 aircraft at Naha Airport. In that case, which was very similar to the Tenerife accident, while the aversion manoeuvre by the All-Nippon Airways plane was of course appropriate, if the two planes had both been jumbo jets, a major disaster of the same kind as occurred at Tenerife would probably have occurred even if this same aversion manoeuvre had been performed.

In the moment that the two aircraft were touching, the bottom of the left outer engine (No. 1 engine) of the jumbo jet scraped against the right wing tip of the MU-2 and the fuel tank that is attached there. Although dents resulted in the outer plate on the bottom of the No. 1 engine cowling of the jumbo jet, the damage was so slight that the crew of that plane did not feel the impact at all (see Fig. 12).

Given the amount of fuel on board and the aerodynamic lift in effect at the time, the height of the bottom of the No. 1 engine above the ground was 6.6 ft, the same as the height of the wing tip of the MU-2. Nevertheless, the damage was limited to grazing the wing tip and fuel tank of the MU-2.

If the jumbo jet had been carrying a large amount of fuel or had been moving more slowly, the wings would have been drooping lower owing to the increased weight and less lift, and it can be inferred that the fuel tank of the MU-2 would have been sucked into the No. 1 engine of the jumbo jet, resulting in a major disaster.

Also, taking one example from the 430 incidents studied, it was a

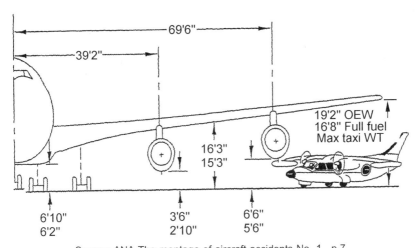

Source: ANA The montage of aircraft accidents No. 1. p.7
Fig. 12 B-747 No. 1 engine and height of MU-2 wingtip

matter of good fortune that the number of passengers aboard numbered only in the several tens so that the aircraft was relatively light when it took off with ice build-up on the wings and experienced reduced lift and several bouts of buffeting, with a resultant 'serious' degree of danger. The plane had travelled over more than half the runway during take-off and the pilot was aware that the 'rotation speed' V_R (the speed at which the nose of the aircraft begins to lift upwards during take-off) had been exceeded, but at that stage there was no choice but to continue with the take-off. If the take-off had been aborted (reject take-off) at that time, the plane would certainly have overshot the runway.

Although the take-off in this incident was, in effect, successful (lift-off occurred just before the end of the runway), it was merely a matter of 'good fortune' that the plane had been light enough to make the take-off possible. For the pilot, it was a severe situation in which there was no room for choice and he was fully aware of the possibility that, if the plane were heavy, the result would be a total-loss accident.

In the same way that danger factors accidentally chain together in the process by which unsafe events occur, there are also cases in which good fortune can be considered as the only reason that a situation did not develop into an accident. Although these cases are attributed to 'good fortune', these cases differ from those in which the reason for recovery cannot be determined from the incident reports.

The cases to which the 'good fortune' reason for recovery applies, at 47 cases or 15.1 per cent, are the next most numerous after 'noticing and acting on one's own mistake' without outside visual information or instruments (at 62 cases or 19.9 per cent).

Including the cases in which the person noticed his own error with the help of outside visual information or information from instruments or abnormal sounds or vibration, the cases in which the person himself took action to recover from the mistake numbered 92, or 29.5 per cent; the cases in which the recovery was due to advice or action by other crew cockpit members numbered 116, or 37.2 per cent. Thus, the incidents in which the recovery was accomplished by the cockpit crew members total 208, or 66.7 per cent of all of the incidents that were studied. The cases in which recovery was due to 'good fortune' is nearly one-quarter of that total.

This fact contains an important problem that cannot be ignored. Although these incidents did not develop into accidents through accidental good fortune, the possibility of their developing into an accident was nevertheless extremely high. Accordingly, persons concerned with aviation should take a serious view of the results of analysing the factors of incidents whose reason for recovery was 'good fortune', and must recognize the need earnestly to deal with the results and urgently devise appropriate improvements and countermeasures.

2.23 Improvements and countermeasures uncovered by analysis

As we have seen so far, there are various factors that affect safety concerning the operation of an aircraft, and in many cases there are multiple background factors that lie behind the occurrence of an incident and lead to human errors. As cited frequently in this book, each of the background factors that have been found from information on 430 incidents that involve the job of flying an airplane is often also one of the background factors of other accidents that had occurred previously.

What is important in terms of preventing accidents, however, is not simply the determination of what background factors there are, but rather how the factors that are involved in the incident are interrelated (i.e. their correlations) at the time of the investigation. Examples of these background factors are the error mode, the reason for recovery, the degree of danger, the phase of information processing in which the incident was determined, the flight phase, and the time zone in which the incident occurred. Such factors exclude those that are considered to be the specific attributes of each incident, such as age at the time of occurrence, total flight hours, type of aircraft, the season in which the incident occurred, the airline company to which the pilot belongs, and occupational classification. In addition, one must take into account the understanding of the actual situation of the danger of each factor and the study and implementation of improvements and countermeasures based on that understanding, as has already been described earlier in this book.

If we first group factors that are strongly related on the basis of the results of a general, multidimensional analysis by applying quantification method III to the 98 factors that were identified from individual analyses of the 430 reported incidents, we will be able to identify classes (clusters, in the terminology of statistics) of incidents that are typical at the time of the investigation.

If we then assign to each of those clusters the name of the factor that characterizes it, we would have clusters for near misses, abnormal atmospheric pressure, cumulonimbus, wind shear, weather observation and reporting system, no ILS, company policies, snow, marking, forgetting, excessive focus, pilot in training/check flight, unexpected changes, fatigue, charts/manuals, parallel runways, malfunctions/ mechanical problems, and visibility hindrances (see Fig. 13).

The near-miss cluster, for example, comprises the factors of other nearby aircraft, improper air traffic control instruction or information, in clouds, and 'averting operation of the other aircraft' as the reason for recovery. The degree of danger is close to 'serious'.

In the Tenerife accident, as described in Chapter 1, the collision occurred because the aversion manoeuvre by the Pan American Airlines (PAA) plane was too late, but the other nearby aircraft, improper instruction or information from air traffic control, and in clouds/fog were contributing factors. Of course, the Tenerife accident was not included in this incident analysis, but this kind of unsafe event would probably not have taken place if the air traffic control instructions and information had been appropriate; even granting that the two planes might have come abnormally close to each other, a collision could have been averted.

The abnormal atmospheric pressure cluster and the cumulonimbus cluster lie near the near-miss cluster, indicating that the probability of a near miss is high under these conditions. This is because, in the case of abnormal atmospheric pressure, an error in setting the atmospheric pressure value results in a high probability of a near miss in relation to the altitude of the aircraft, while, in the case of cumulonimbus, aircraft flying outside the air ways to avoid those clouds results in a high

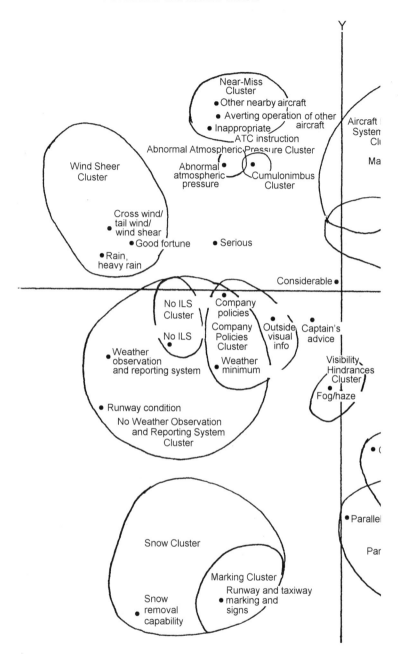

Fig. 13 Cluster of pilot-related human factor incidents (the *X–Y* axis plane) (see colour plate section)

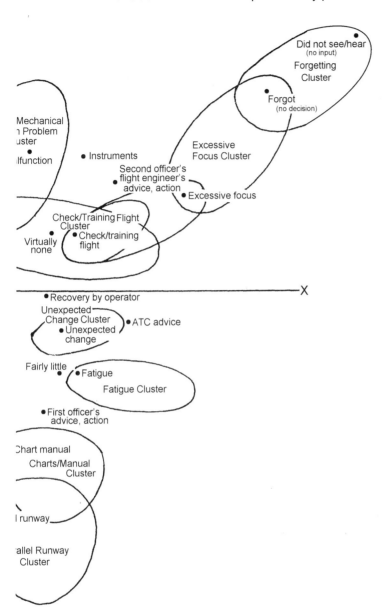

probability of a near miss in relation to the horizontal plane because the aircraft may be flying at the same altitude.

It can be said that an effective way to prevent this kind of unsafe event represented by each cluster of factors is to study and implement improvements and countermeasures for the most important of the factors that make up that cluster.

The main improvements and countermeasures to be proposed on the basis of that study, with detailed descriptions reluctantly omitted, are: proper advice by air traffic control concerning other nearby aircraft; the installation of wind shear monitoring equipment; the installation of an instrument landing system; an expanded weather observation and reporting system; the establishment of a 'safety first' policy by airline companies; securing of snow and ice removal capabilities for aircraft and snow removal capabilities for runways and taxiways; improved taxiway and spot markings; clarification of the division of duties of crew members; the presence of a monitoring pilot during check and training flights; proper advice by air traffic control when unexpected changes are made; improvement in the flight duty schedule for long-distance flights; preparation and improvement of manuals; and feedback on specific incidents.

The most difficult cluster to deal with, however, is the 'forgot' cluster. That cluster consists of 'did not see' or 'did not hear' in the information receiving phase of the error mode, 'forgot' in the decision phase, and 'forgot' in the action (operation) and instruction phase, and no factors that correlate strongly with others are found. If we were determined to derive such a correlation, we would find it in the factors 'light workload' and 'cruising'. That fact indicates that forgetting easily occurs when there is less than a moderate amount of stress or stimulation.

Furthermore, 24 of the 430 reported incidents involve such so-called simple forgetting, a fact that increases the credibility of the investigations and points to the great difficulty of finding ways to prevent forgetting.

An example is forgetting to return the INS mode selector switch to the proper position after a manoeuvre to avoid a cumulonimbus. This mode selector switch – a switch that selects the input to the automatic control system, including the INS mode, heading mode, and approach mode – is positioned on the control panel directly in front of the pilot's seat, where it is easily seen. Switching between the modes that are most important during flight, however, is accomplished by turning a single knob by a mere 15° or so (see Fig. 14).

Thus, although the setting operation is extremely simple, and in spite

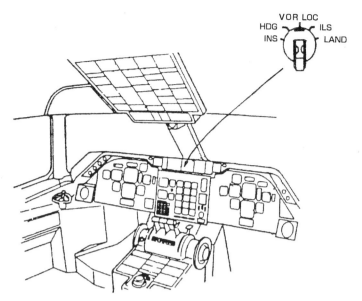

Fig. 14 INS mode selector (B-747)

of the fact that it is such an important operation, the letters that indicate the switch position are small and the resistance felt when turning the knob to set the mode is light, so it is difficult to remember the mode in which the plane is currently flying. For that reason, it may be difficult to recall that fact in the event that one has forgotten to perform the operation.

Of course, it is possible to know the navigation situation from other navigation display devices, but, in situations where something has been forgotten, even the confirmation of that operation is also forgotten. Such is the nature of forgetting, and even audio warning devices have inherent limits. The setting of a standard for the point in time at which a warning device is to be sounded is difficult, and there are reports of cases in which a person in the state of having forgotten something did not even hear the sound of an alarm that should have gone off.

The incidents for which the operation of warning devices was the reason for recovery number only 8, or no more than 2.5 per cent of the 430 reported incidents, which may indicate the difficulty of catching a person's attention with warning devices. Forgetfulness is one of man's greatest weaknesses, and there is probably a need to gather the wisdom of various intellectual disciplines to find ways to solve the problem of human forgetfulness (methods of effectively drawing attention to problems).

2.24 Degree of danger in 'hard factors' and 'soft factors'

When viewed comprehensively by means of quantification method III, the various latent danger factors that affect the operation of commercial airliners are broadly classified into two categories according to their nature, and their contrasting differences in degree of danger and reason for recovery can be understood.

As described in Chapter 1, if the three-dimensional stereoscopic graph (see Fig. 15) is viewed from the front (the $X-Y$ axis plane), as shown in Fig. 16, all of the meteorological factors and the factors related to navigational aids and airport safety facilities are found on the left side of the X axis. It can be seen that these are closely related to 'vacillation'

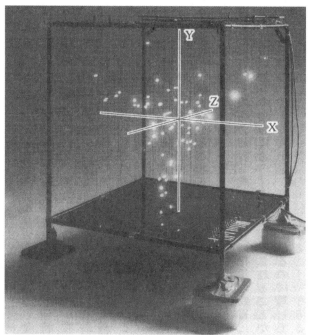

Fig. 15 A model for representing the stereoscopic 3-D graph of quantification method III (the X, Y, and Z axes were added in the printing process)
As this photo illustrates, the positions of the various factors are fixed in a Euclidean space. The positions of the factors shown in Figs 16, 18, and 19 appear different at first glance, but in these cases it is the viewer's perspective that is different, and not the positions of the factors themselves. This is much the same as our view of the positioning of the stars: the arrangement of stars in a constellation as seen from Earth will be quite different from the apparent arrangement in the same constellation as seen from a different position in space.

and 'could not see or could not see clearly/could not hear or could not hear clearly' in the information receiving phase of the error mode, to 'vacillation' and 'unreasonableness or high-handedness' in the decision phase, and to 'recklessness', 'forcefulness', 'excessive action/operation/work', or 'difficult action or instruction/inability to act or work' in the action (operation) or instruction phase.

On the right side of the X axis, in relative symmetry to this, lie a variety of factors such as pilot in training/check flight, unexpected changes, fatigue, misplaced deference or excessive concern, impatience or fluster, interdependency, long-distance flights, and heavy or light workload. The error modes that are close to these are: 'did not see or did not hear', 'failure to see or hear', 'saw incorrectly', excessive focus, preconception, and delay in the information receiving phase; forgot, wrong conviction/hasty conclusion/one's own interpretation, and delay in the decision phase; and forgot, incorrect action/instruction or mistaken/inappropriate action or instruction, and delay in the action (operation) or instruction phase. From that fact we can see that the relation to the factors described above is strong.

For convenience in explanation, we will refer to the various factors that are distributed on the left side of the X axis as the hard factor group and those on the right side as the soft factor group.

The degree of danger of the hard factor group is high ('serious') and the reason for recovery is 'good fortune' rather than human action or advice. For the soft factor group, on the other hand, the degree of danger is low ('fairly little' or 'virtually none'). As for the reason for recovery, from the strong relationship with advice by the flight engineer, instruction or advice by the air traffic controller, advice or action by the copilot, and so on, we can see that the danger was averted by the action or advice of a human. While the information processing phase that determines the incident is the information source phase in the case of the hard factors, it is the information receiving phase in the case of the soft factors.

That is to say, although the background factors of both of the two factor groups are the same in that they lead to human error, the hard factor group comprises non-human factors (weather conditions, navigational aids, and airport safety facilities) that involve an extremely high degree of danger and easily lead to errors of 'could not see', 'could not see clearly', vacillation, unreasonableness, recklessness, and difficult action or instruction/inability to act or work, and are unable clearly to display the normal capabilities possessed by humans.

In contrast to that, the soft factor group comprises factors that have a

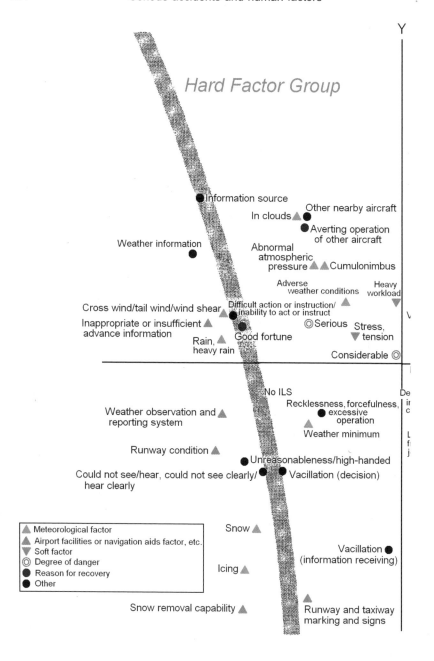

Fig. 16 The 3-D stereoscopic graph viewed from the front side (the *X–Y* axis plane), showing the relations of the hard and soft factors and the error mode and degree of danger (see colour plate section)

Soft Factor Group

Did not see/hear (no input) ●

Forgot (no action/instruction) ● ● Forgot (no decisions)

● Second officer's, flight engineer's advice, action
● Excessive focus

Virtually none ◎ ▼ Check/training flight

Impatience/
fluster   ▼ Excessive concern/misplaced deference
▼   ● Delay (action/instruction)
———————————————— ● Failure to see/hear/notice ——————————————— X
Recovery by
operator

● ATC advice

Delay (decision)/ ● ● Unexpected change in ATC clearance, sudden occurrence
Insufficient
consideration ● Delay (information receiving)

Fairly little ● ▼ Fatigue/lack of sleep
Long-distance flight/
frequent take-off and landing, ▼ ▼ Over-reliance interdependence/
jetlag, low remaining fuel   excessive familiarity among crew
● First officer's advice, action
● Incorrect action/instruction, inappropriate action/instruction
● Information receiving

Wrong conviction/hasty conclusion/one's own interpretation
● ● Preconception

● Saw incorrectly/heard incorrectly/
mistaken impression

√eigenvalue
1st dimension: 0.6090
2nd dimension: 0.5572
3rd dimention: 0.5449

relatively low degree of danger and lead to errors such as 'did not see' or 'did not hear', 'saw or heard incorrectly', delay, excessive focus, and preconception in the information receiving phase, wrong conviction and delay in the decision phase, and 'forgot', incorrect action, and delay in the action (operation) or instruction phase, errors which have their origins in the natural weaknesses of humans.

From these facts we can understand that, for the hard factor group, improvements in hardware systems such as the establishment of weather observation and reporting systems and supplementation of navigational aids and airport safety facilities are essential, and without such improvements we cannot expect to prevent these kinds of incident from developing into accidents by human advice or instruction. For the soft factor group we can understand that the development of an incident into an accident can be prevented if the chain of danger factors is broken by appropriate human advice or instruction.

Factors that concern the attributes of an incident include the airline company to which the reporter of the information belongs, his age, his total flying hours and position, the season in which the incident occurred, the person's position at the time of the incident, and the type of aircraft. If we superimpose the results of a quantification method III analysis of the factors on the results of a comprehensive analysis of the factors related to the incident, we see that captain pilots who are at least 50 years old, flying mainly local routes in Japan, and have over 10 000 total flying hours correspond to the hard factors that are strongly related to the 'serious' degree of danger and the 'good fortune' reason for recovery (see Fig. 17).

This fact, too, indicates that the hardware improvements mentioned above are essential for coping with the hard factor group, and we cannot look to pilot experience and skill as the solution to those problems.

2.25 Urgently needed response to 'hard factors'

If we now view the three-dimensional stereoscopic graph from the top (the X–Z plane, see Fig. 18), we can understand the following. Firstly, let us look at the relation of the latent danger factors of the hard factor group and 'could not see or could not see clearly' and 'vacillation' of the error mode information receiving phase, 'vacillation' and 'unreasonableness or high-handedness' of the decision phase, and 'recklessness, forcefulness or excessive action/operation/work' and 'difficult action or instruction/inability to act or work' of the action (operation) or

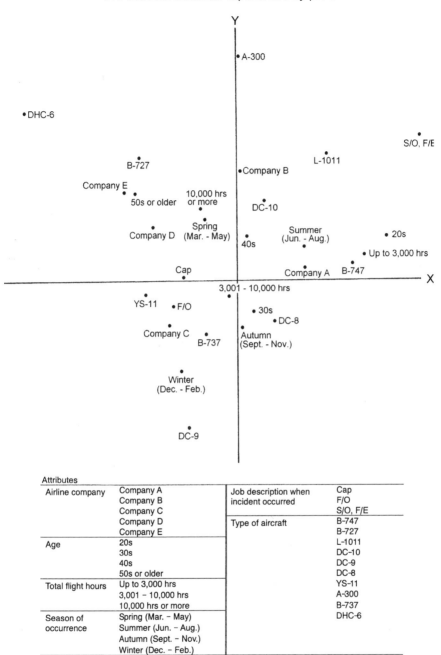

Fig. 17 Attributes of pilot-related incidents (the X–Y axis plane)

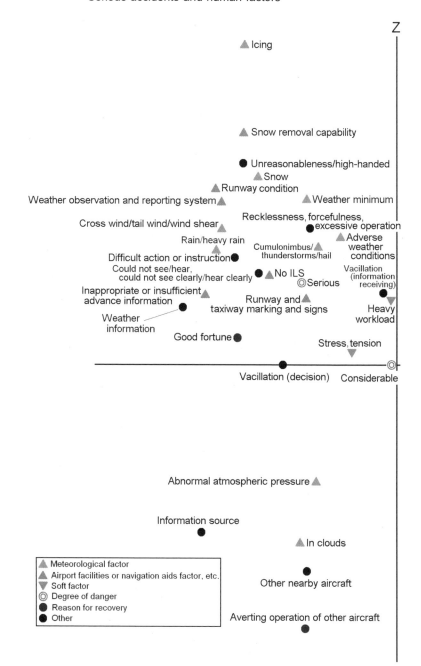

Fig. 18 The 3-D stereoscopic graph viewed from the top (the *X–Z* axis plane) (see colour plate section)

Excessive focus
●

Delay (action/instruction)
●

Forgot
● (no decision)

Did not see/hear
● (no input)

● Delay (decision)/insufficient
consideration

● Forgot
(no action/instruction)

Impatience/fluster
▼

● Delay (information receiving)

▼ Check/training flight

Recovery by
operator ● ▼ Excessive concern/misplaced deference
First officer's advice, action
● ● Second officer's, flight engineer's advice, action
──────────────────────────────●Failure to see/hear/notice─────────────── X
▼ Unexpected ▼ Over-reliance interdependence/
change in ATC excessive familiarity among crew

Virtually none ◎

Fairly little ◎ ▼ Fatigue

Long-distance flight/frequent take-off and
Information landing, jetlag, low remaining fuel
receiving ● ▼

● Incorrect action/instruction, inappropriate action/instruction
● Preconception
● ATC advice

● Wrong conviction/hasty conclusion/one's own interpretation
●
Saw or heard incorrectly/mistaken impression

instruction phase. We can see that all of the meteorological factors other than abnormal atmospheric pressure and clouds, which are intimately related to near misses, and all of the factors that are related to navigational aids and airport safety facilities, including weather observation and reporting systems, have a deep relationship with these error modes. We can further see that the degree of danger of these factors is 'serious' and the reason for recovery is accidental 'good fortune'.

Looking at the stereoscopic graph from the side (the Y–Z plane, see Fig. 19) too, this relationship does not change. We can see that the hard factor group danger factors mentioned above have a deep relationship with 'could not see or could not see clearly' of the error mode information receiving phase, with 'vacillation' and 'unreasonableness or high-handedness' of the decision phase, and with 'recklessness, forcefulness, or excessive action/operation/work' and 'difficult action or instruction/inability to act or work' of the action (operation) or instruction phase. We can further see that the degree of danger of these factors is also 'serious' and the reason for recovery is also accidental 'good fortune'.

The fact that the degree of danger for the hard factors is thus extremely high in all three dimensions shows that, for safety in aviation, the establishment of improvements and countermeasures for these hard factors is an urgent task.

As for these hardware improvements, under the Civil Aeronautics Law of Japan, the responsibility for airport installations and management is entrusted entirely to the Minister of Land, Infrastructure and Transport and the local public entity, public corporation (the New Tokyo International Airport), or special corporation (Kansai International Airport), according to the type of airport. Therefore, this problem cannot be solved as an internal problem of one airline company; it is a problem that must be addressed as a governmental administrative matter.

Although there is steady ongoing improvement in airport and aviation safety facilities, we see little improvement in the weather observation and reporting systems. Although counter to the age of corporate restructuring, increasing the number of weather observation personnel is an indispensable measure for ensuring aviation safety. Even with today's highly developed navigation technology, the most recent and highly accurate weather information greatly affects air operations.

The incidents for which the fact that weather information differed

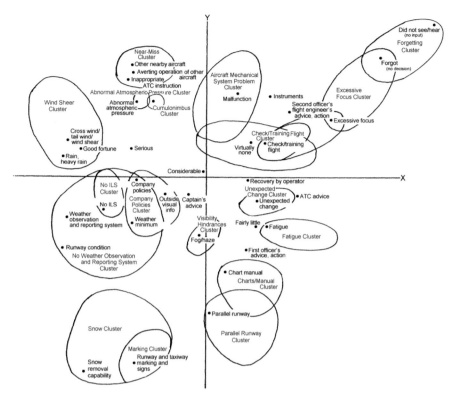

**Fig. 13 Cluster of pilot-related human factor incidents
(the X-Y axis plane)**

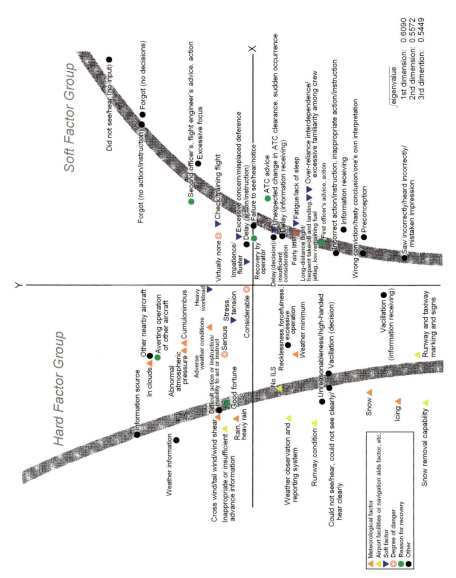

Fig. 16 The 3-D stereoscopic graph viewed from the front side (The X-Y axis plane), showing the relations of the hard and soft factors and the error mode and degree of danger

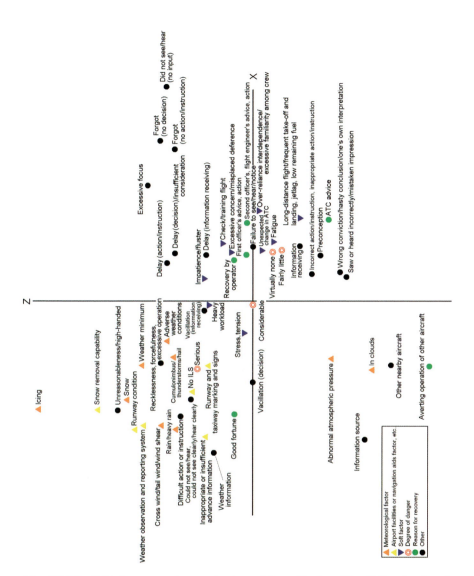

Fig. 18 The 3-D stereoscopic graph viewed from the top (the X-Z axis plane)

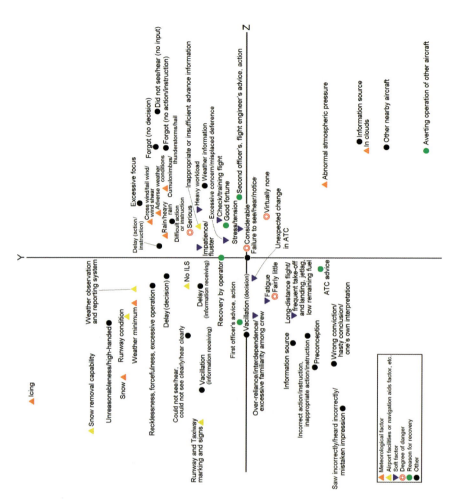

Fig. 19 The 3-D stereoscopic graph viewed from the side (the *Y–Z* axis plane)

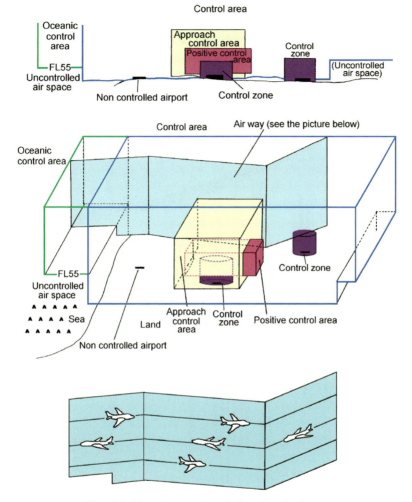

Fig. 20 Air space concept drawing in Japan

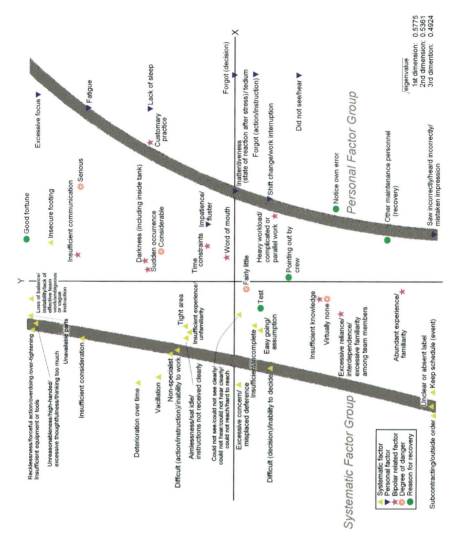

Fig. 22 Two major characteristics and bipolar related factor 'occurrence' incidents. Aircraft maintenance personnel and latent danger in aircraft

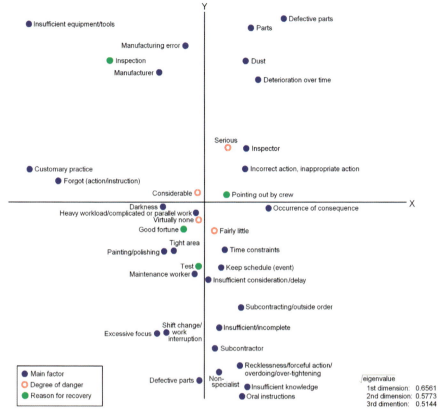

**Fig. 23 Distribution of main factors for 'discovery' incidents.
Aircraft maintenance personnel and latent danger in aircraft**

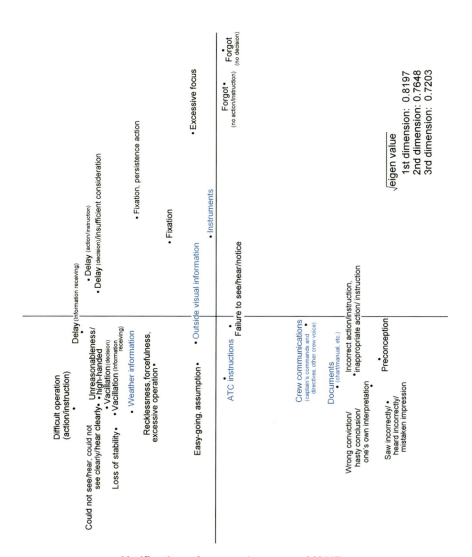

Verification of appropriateness of MAIR

√eigen value

1st dimension: 0.8197
2nd dimension: 0.7648
3rd dimension: 0.7203

Difficult operation
(action/instruction)

Delay (Information receiving)

• Delay (action/instruction)
• Delay (decision)/insufficient consideration

Could not see/hear, could not
see clearly/hear clearly • • Unreasonableness /
• high-handed

Loss of stability •   • Vacillation (decision)
• Vacillation (Information
receiving)

• Vacillation (Information
receiving)

• Weather information

• Fixation, persistence action

Recklessness, forcefulness,
excessive operation •

• Fixation

• Excessive focus

• Outside visual information

Easy-going, assumption •

• Instruments

Forgot •
(no action/instruction)

Forgot •
(no decision)

ATC instructions •   •
Failure to see/hear/notice

Crew communications •
(captain's commands and
directives, other crew voice)

Documents •
(chart/manual, etc.)

Incorrect action/instruction,
• inappropriate action/ instruction

Preconception •

Wrong conviction/
hasty conclusion/
one's own interpretation

Saw incorrectly/ •
heard incorrectly/
mistaken impression

from the actual weather was a contributing cause numbered 20 out of the 430 reported incidents. The degree of danger associated with those 20 incidents was quite high, with ten having a 'serious' degree of danger, eight having a 'considerable' degree of danger, and two having a 'fairly little' degree of danger.

According to the Civil Aeronautics Law (Article 99), the Minister of Land, Infrastructure and Transport must provide the crew of an aircraft with the information that is required in the operation of the aircraft, and weather information is part of such required information (Article 209-(2), Para. 1, Subpara. 6 of the Civil Aeronautics Regulations of Japan). Furthermore, the captain of the aircraft must, before departure, check the weather information for the departure area, the air route, and the destination area (Article 73-(2) of the Civil Aeronautics Law of Japan and Article 164-(14), Para. 1, Subpara. 4 of the Civil Aeronautics Regulations of Japan).

However, at the end of March 1988, after the completion of this research, there were as many as 11 airports, including Misawa, for which formal weather information for the destination could not be obtained before an early morning departure because the weather observation at the destination did not begin before the departure time. When flying with one of these airports as the destination, without being able to obtain a formal weather observation report, there is, in practice, no recourse other than to obtain a rough account of the local weather from a member of the company's staff at the destination airport.

2.26 Danger factors increasing in importance and targeted for improvement

Next, if we look at the degree of danger of the soft factors, viewing the three-dimensional stereoscopic graph from the front (the $X-Y$ axis plane, Fig. 16) and from the top (the $X-Z$ axis plane, Fig. 18), we cannot find soft factors that are associated with the 'serious' degree of danger or the 'good fortune' reason for recovery.

Viewing the three-dimensional graph from the side (the $Y-Z$ axis plane, Fig. 19), however, the soft factors impatience/fluster, heavy workload, check/training flight, unfamiliarity, meddlesomeness, stress/tension, misplaced deference/excessive concern, and scheduling delay/diversion, the information receiving phase error mode factors 'did not see or did not hear', 'excessive focus', and 'delay', the decision phase factors 'forgot', 'easy-goingness/assumption', and 'delay', and the action (operation)/instruction phase factors 'forgot', 'delay', and

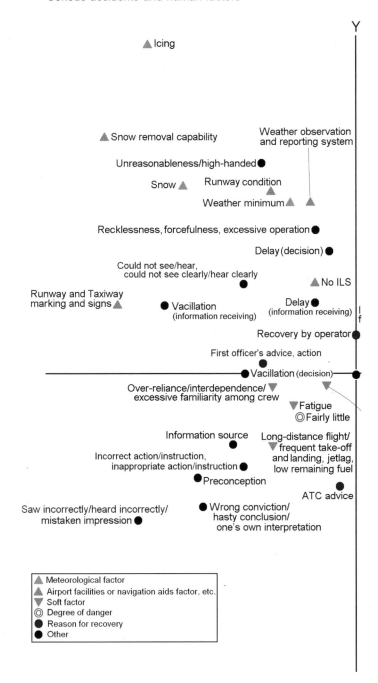

Fig. 19 The 3-D stereoscopic graph viewed from the side (the Y–Z axis plane) (see colour plate section)

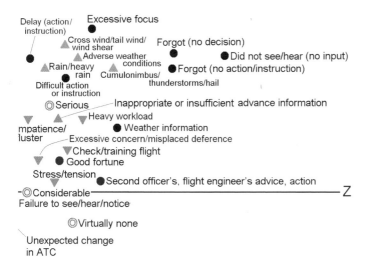

Delay (action/ instruction)
Excessive focus
Cross wind/tail wind/ wind shear
Forgot (no decision)
Adverse weather conditions
Did not see/hear (no input)
Rain/heavy rain
Cumulonimbus/ thunderstorms/hail
Forgot (no action/instruction)
Difficult action or instruction
◎ Serious
Inappropriate or insufficient advance information
Heavy workload
mpatience/ luster
● Weather information
Excessive concern/misplaced deference
▼ Check/training flight
● Good fortune
Stress/tension
● Second officer's, flight engineer's advice, action
-◎ Considerable ————————————————————————— Z
Failure to see/hear/notice

◎ Virtually none

Unexpected change in ATC

▲ Abnormal atmospheric pressure

● Information source
▲ In clouds

● Other nearby aircraft

● Averting operation of other aircraft

'fixation' appear near the 'serious' degree of danger and the 'good fortune' reason for recovery.

This indicates that pilots easily fall into states of impatience/fluster, heavy workload, unfamiliarity, meddlesomeness, stress/tension and misplaced deference or excessive concern during check and training flights and scheduling delays or diversions (destination change), and, in such cases, there is a tendency to make the errors of excessive focus, fixation, easy-goingness/assumption, delayed behaviour, and forgetting.

While the instructor captain is conducting the training during a training exercise, he must also perform what are normally the duties of the copilot, to which he may be unaccustomed. In this situation, however, it often happens that the captain falls into excessive focus on the training and forgets to perform the copilot's task of communicating with air traffic control and lands the plane without clearance, forgets to lower the landing gear, or tends to display meddlesomeness with respect to the trainee, and the instructor's workload is heavy. The trainee pilot is in a state of unfamiliarity and easily falls into impatience or fluster or forgetful behaviour.

In the case of check flights, too, meddlesomeness creates excessive tension in the pilot who is being examined. Thus, even if the flight engineer notices a problem, misplaced deference or excessive concern may prevent him from offering advice to the pilot.

Analysis of this dimension reveals that the factors associated with pilot in training/check flight and arrival delay/diversion are related to the 'serious' degree of danger. This suggests that, when hardware improvements are made in Japanese commercial aviation, the factors that surround the pilot in training/check flight and scheduling delay or diversion will warrant close examination as factors associated with a high degree of danger.

Many unsafe events in the check or training flight are averted as the result of advice from the flight engineer. Now, many aircraft have only two persons in the cockpit. A countermeasure for that is to have a third person, a monitor pilot, in the cockpit. By doing so, it is possible to prevent errors such as excessive focus, scheduling delay/diversion, or forgetting, which the inspecting or training captain and pilot trainee are prone to make during check or training flights.

By viewing the results of analysing individual incidents three-dimensionally with quantification method III in this way, relationships that cannot be understood from the individual analyses become clear.

Chapter 3

Imminent danger experienced by air traffic controllers

3.1 Air traffic control space and various air traffic control services

After World War II, the skies that were previously open to free, unhindered flight became crowded as commercial airlines developed, and, furthermore, as propeller aircraft were replaced by jet aircraft, aircraft speed increased and the danger of air collisions and interruptions in the flow of air traffic increased. The air traffic control system was created to maintain the appropriate distance between aircraft in the crowded skies and avert the danger of collisions. Its purposes also include maintaining and promoting orderly, regular air traffic.

It is recognized that the worldwide uniform system for air traffic control created by the International Civil Aviation Organization (ICAO), a special organ of the United Nations, is indispensable to the development of air transportation and to the securing of safety in the field. Based on this awareness, Appendix 11 of the International Civil Aviation Convention was written and set as an international standard and recommendation system, and all signatory countries were requested to implement it.

The air traffic service involves the following three types of work:

1. *Flight information service.* Providing an aircraft in flight with information about the circumstances of the airspace and weather information for the region around the aircraft.
2. *Alerting service.* Emergency tasks for helping aircraft that require search and rescue support.
3. *Air traffic control service.* Giving traffic control approval (clearance by the air traffic controller concerning matters of the route and altitude for aircraft that will enter the controlled airspace under instrument flight rules) and instructions.

Wherever on the globe an aircraft is flying, it can receive the benefit of the flight information and alerting service. In contrast to this, the air traffic control service is performed for aircraft flying in three controlled airspaces: control area, control zone, and oceanic control area.

The scope of the control area and the control zone (i.e. the airspace) is divided into plane and three-dimensional sectors.

Annex 13 of the Convention on International Civil Aviation (hereafter referred to as the Chicago Convention) states that air traffic services airspace shall be categorized into classes from A to G (Annex 13, Chapter Article 2, 6), and defines the content of air traffic services provided by IFR and VFR devices for each limited-range airspace (Annex 13, Appendix A).

According to the Chicago Convention, Article 38:

Any state that finds it impracticable to comply in all respects with any such international standard or procedure, or to bring its own regulations or practices into full accord with any international standard or procedure after amendment of the latter, or that deems it necessary to adopt regulations or practices differing in any particular respect from those established by an international standard, shall give immediate notification to the International Civil Aviation Organization of the differences between its own practice and that established by the international standard.

In the case of Japan, however, although the Minister of Land, Infrastructure and Transport has given public notification of items related to airspace that differ from international standards, as of this writing (May 2003) ICAO has not yet been officially notified of these differences.

It would appear that, not only in Japan but in other countries as well, air traffic services, flight rules, and the scope of control areas and control zones do not necessarily conform to ICAO standards.

Nevertheless, all human incidents related to air traffic services as described in this chapter are based on reports from air traffic controllers in Japan. We are thus able to make some general observations regarding the most significant differences between Japanese and international standards with regard to airspace, as well as the unique characteristics of the Japanese view of this subject.

In Japan, aircraft flying by VFR in an approach area are not obligated to communicate with air traffic control units. Among the airspace classes defined as standards by ICAO, however, there are airspace classes (B to D) considered equivalent to approach areas in

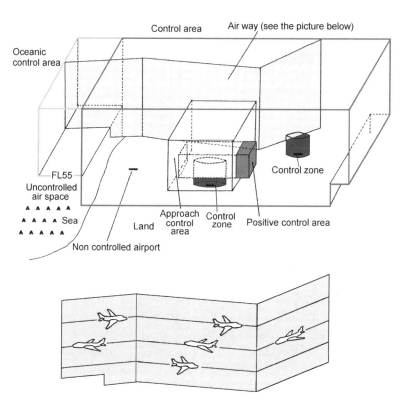

Fig. 20 Airspace concept drawing in Japan (see colour plate section)

Japan in which aircraft flying by VFR are obligated to establish such communications.

Furthermore, in Japan, those areas with a specific, limited scope near airports with high traffic volumes are defined as Positive Control Areas (PCAs) (see Fig. 20), where aircraft cannot fly using VFR without authorization from the control units. In other words, in PCA airspace, aircraft must fly using instrument flight rules.

The oceanic control area in Japan is the oversea airspace within the flight information region for which Japan is responsible (according to the Convention on International Civil Aviation, each signatory country is assigned an air traffic controlled airspace).

Because the air traffic control service in Japan is under the authority of the Minister of Land, Infrastructure and Transport (although some responsibility is delegated to the Director of the Defence Agency), air traffic controllers are all government employees, and instructions from air traffic controllers, by law, represent instructions from the Minister of Land, Infrastructure and Transport.

In order to ensure a clearer understanding, the specific examples presented in this chapter will be based on the conditions in Japan as described above. Explained in terms of their relationship to the controlled airspaces, the tasks of air traffic control services are as follows.

The control zone is a cylindrical airspace that usually extends in a 5.4 mile radius around the marked point of the airport (a point on the ground roughly at the centre of an airport, expressed as a latitude and longitude) and to a height of 3000 ft above the ground surface.

Within that airspace, air traffic control performs tower control services, including the issuing of instructions concerning taxiways to aircraft moving on the ground and the granting of clearance for take-off or landing to aircraft that are to take off or land. This is accomplished through visual observation from the airport tower, and is usually referred to as tower control.

The control area is the broad expanse of airspace within the controlled airspace other than the control zone. It extends from a minimum altitude of 650 ft above the ground surface and has no particular upper altitude limit. In this airspace, air traffic control service involves mainly supervision of aircraft that are cruising in the air route and is usually referred to as 'en route' control. The other aspect of control service is directed at aircraft that are flying in the oceanic control area and those that are entering or leaving airports that do not perform approach control.

Within the control area, there are approach areas, which are designated airspaces around major airports in which the air traffic is dense. These approach areas are located in the middle between the control zone of an airport and the control area in which there are air routes. Its shape varies with the airport, but it extends to a radius of about 42 miles from the airport marked point and to an altitude of about 15 000 ft. In the approach area, air traffic control service is provided for aircraft that are approaching the airport, departing from it, or transiting it in this airspace.

That service involves three missions. One is the approach control service for aircraft that are climbing after take-off, descending for landing, and passing through this airspace. Another is terminal radar control, which employs an Automated Radar Terminal System (ARTS) and a Terminal Radar Alphanumeric Display System (TRAD). The third mission is ground-controlled approach, in which aircraft are guided up to a point near the runway by using Air Surveillance Radar (ASR) and Precision Approach Radar (PAR). Of these, the approach control service and terminal radar control service are usually referred to as approach.

In Japan today, most air traffic control areas are covered by Air Route Surveillance Radar (ARSR), but, because the range of one ARSR is a radius of only 200 nautical miles, oceanic control is most often done without radar. For that reason, oceanic air traffic control involves the use of an operations table (a strip that contains information such as aircraft call signs, type, speed, route, and altitude, and air traffic controllers record the instructions they have given, changes, and so on sequentially as they work), which is automatically generated by a computer (FDP, the flight plan data processing system) on the basis of the flight plan and information obtained directly by communication with the aircraft or relayed by the air traffic communications specialist.

3.2 Flight rules: Visual Flight Rule (VFR) and Instrument Flight Rule (IFR) aircraft

Aircraft flight rules include visual flight rules and instrument flight rules.

Visual flight rules are the method of flying used when weather conditions (visual meteorological conditions) permit the pilot to see other aircraft and obstacles by eye, and so maintain a visual separation and avoid collisions. Visual meteorological conditions are conditions under which it is possible for the pilot to avoid obstacles by looking out the cockpit windows, without relying on instruments alone, and are set specifically on the basis of visibility, distance from clouds, and altitude of clouds, according to the airport vicinity or air route. Usually it is required that the flight visibility is 16 000 ft or more and there are no clouds around aircraft. A plane flying under visual flight rules is usually referred to as a VFR plane.

Instrument flight rules, on the other hand, are a method of flying in which, for flying routes, etc., the plane is operated entirely according to instructions from the air traffic controller. Unless specifically instructed to maintain visual separation, the responsibility for collision prevention lies primarily with the air traffic controller. Even when flying under instrument flight rules, however, the pilot must maintain a visual lookout and not rely only on the instruments when visual meteorological conditions exist (Article 71-2 of the Civil Aeronautics Law of Japan). A plane that is flying under instrument flight rules is usually referred to as an IFR plane.

Commercial airliners must be flown under instrument flight rules, except for extremely limited cases such as local flights for training, inspection, or special invitation, and search and rescue flights. Thus, pilots of aircraft flying under instrument flight rules are obligated to fly according to the instructions of air traffic control.

Aircraft other than civil commercial airliners, which are referred to as light aircraft, including business aircraft, private planes, military aircraft, etc., normally fly under visual flight rules when visual meteorological conditions exist, and are not always obligated to follow the instructions of air traffic control. Accordingly, the aircraft that are flying within air traffic control areas and terminal control zones include both IFR planes, which must always follow the instructions of air traffic control, and VFR planes.

Control areas above the area around airports where traffic is heavy, however, have been designated by the notification of the Minister of Land, Infrastructure and Transport as positive control areas, and it is stipulated that aircraft must fly under instrument flight rules within those airspaces.

In the case of instrument meteorological conditions (poor weather conditions determined by visibility and cloud conditions), light aircraft also must fly under instrument flight rules in control areas and control zones, and thus must always fly according to the instructions of air traffic control. Such light aircraft are, in principle, not allowed to fly in other airspaces under instrument meteorological conditions.

Because civilian light aircraft usually fly under visual flight rules, pilots of those planes are unfamiliar with communicating with air traffic controllers. In cases where the weather deteriorates and the flight rules switch to instrument flight rules, unlike the case with pilots of commercial airliners, such unfamiliarity often results in perplexity when communicating with the air traffic controller.

Communication between air traffic control and aircraft is done over VHF (very-high-frequency) radio in the case of civilian aircraft, UHF (ultrahigh-frequency) radio in the case of military aircraft, and HF (high-frequency) radio in the case of aircraft flying over the ocean. The frequency used, however, differs with the organization that is in control of the airspace.

3.3 Air traffic control service work flow and work assignment

The work of air traffic control service is performed by a team, and the flying of a civil commercial airliner is performed by a crew. That is to say, both types of work are the same in that the tasks involved are divided up among many persons. The work of operating an airliner, however, is a collaboration towards a single purpose at a single point in time. In contrast to that, the tasks involved in the work of air traffic control are assigned by different units for the different flight phases of the aircraft. Also, even within a single unit, various tasks are performed by different controllers, such as the controller who co-ordinates communication with other units (the co-ordinator position) and the controller who gives instructions and advice to aircraft (the air traffic controller, who monitors and operates radios).

That is to say, the sequence of air traffic control services for multiple aircraft that are making transitions through the various phases of flight, from taxiing on the ground to take-off, climbing, cruising, descent, approach, landing, taxiing on the ground, and parking, is divided up among units and individual positions to be dealt with by a conveyor system in work flow fashion.

Each controller is individually assigned responsibility for air traffic control services within a limited temporal and spatial scope, and decisions on instructions and advice to each aircraft and the co-ordination of communication with other controllers or transfer of control to another controller are generally made by that single individual. Exceptions include training situations and when a controller notices that a neighbouring controller is not handling his own task well and offers advice. Accordingly, controllers normally do not know about the traffic conditions within the scope of responsibility of the controllers seated next to them.

Here, we will have a look at the flow of air traffic control service for a scheduled flight from Haneda Airport to Fukuoka Airport.

Firstly, before starting the aircraft engines, the pilot contacts the clearance delivery controller in the Haneda tower by radio to declare his desired flight route and altitude and request clearance. The Haneda control tower relays that information to the Tokyo area control centre. Upon receiving the information, the Tokyo area control centre selects the optimum flight route and altitude according to the traffic conditions and issues clearance. The pilot receives the clearance from the clearance delivery controller in the control tower and then starts the engines.

When preparation for taxiing has been completed, the pilot calls the ground controller for clearance to proceed to the runway. The ground controller gives instructions for the best taxiways to use to move the aircraft near to the runway and for establishing contact with the tower controller.

When the take-off preparations have been completed, the pilot calls the tower controller for clearance to take-off. The tower controller communicates with the departure controller in the instrument flight control room to confirm that the aircraft can depart, makes a safety check, and then issues clearance for take-off. Then, instructions are given to switch the communication radio over to the frequency of the departure controller promptly after take-off.

Up to this point, the air traffic control work has been tower control service. The airspace in which mainly tower control service is performed in this way is the terminal control area.

Next, the departure controller has the aircraft climb while preventing collisions with other aircraft by means of the set flight route or radar guidance, and, when the aircraft nears the boundary of the area for which he is responsible, transfers control to the Tokyo area control centre, an act referred to as hand-off. This work of the departure controller, and the work of the arrival controller, which is described later, together are referred to as approach control service, and, when performed using radar, the work is called terminal radar control service.

The Tokyo area control centre is mainly responsible for supervising aircraft that are in the cruising stage and are flying the route for which they have been cleared. When a pilot request for an en route change in altitude or route is received, this centre considers the relation of the aircraft to other aircraft and takes appropriate measures. Because the Tokyo area control centre is responsible for a large airspace, that airspace is divided into a number of sectors. Cruising aircraft thus pass through several sectors and must change radio frequencies for each hand-off between sectors.

When the plane passes the vicinity of Okayama, control is handed off to the Fukuoka area control centre. The Fukuoka area control centre responsible hands off to the arrival controller of Fukuoka Airport, the destination airport, so the altitude of the aircraft is lowered to a suitable altitude for that purpose while maintaining the safety of this and other aircraft. Then, as the jurisdictional boundary of the airspaces is neared, control is handed off to the arrival controller of Fukuoka Airport.

The control work up to this point is en route control service. The airspace for performing en route control service is the control area.

The arrival controller, who is in the instrument flight control room of Fukuoka Airport, gradually lowers the altitude of the aircraft while avoiding collisions with other aircraft by means of the standard landing path (a standard approach path that is set for each runway for aircraft that are landing by instrument flight rules) or by radar guidance safely to guide the aircraft to the runway. Also, an order is assigned to the various aircraft that have arrived for landing from many directions at the same time. Those aircraft are placed in the final approach course (final) at a fixed separation and then control is handed off to the tower controller in the tower. The work up to this point is approach control service.

After hand-off, the pilot sets the communication radio to the frequency of the tower controller and contacts the controller for clearance to land. The tower controller checks the runway for safety and then issues clearance for landing. After the plane has landed, the tower controller passes control over to the ground controller, who guides the plane through the taxiways to the parking spot. When the plane reaches the parking spot, all of the air traffic control service for this flight is completed.

3.4 Peculiarity of air traffic control service

Air traffic control service has a remarkable special feature that distinguishes it from every other field of work. This peculiarity is that, throughout the world, information and intentions are conveyed between people by means of radio communication alone in extremely short time periods and using a minimum set of English language terms, even outside areas in which English is spoken.

For that reason, air traffic control is a special domain of work in which human factors are most easily brought into play.

The pilot must understand the intentions of the air traffic controller from a minimum set of English terms and expressions and immediately take appropriate action. If there is difference in awareness between pilot and controller, there is an extremely high probability that an unsafe event will occur immediately.

Even if the air traffic controller is giving appropriate instructions and advice, wrong decisions or actions by the pilot may result because the pilot hears incorrectly or fails to hear, or because of radio problems or other pilot-side problems. Needless to say, controller-side problems such as mistaken or inappropriate instructions due to mistaken impressions, misspeaking, and unclear pronunciation or forgetting to give instructions also occur.

Important items in controller instructions, particularly clearance for take-off or landing, altitudes, routes, and so on, are normally confirmed by pilot-side read-back, in which the pilot repeats the information back to the controller. However, there are also cases of 'wishful hearing', in which the controller fails to notice an error in the read-back because he is expecting to hear the pilot repeat the information that the controller has given. Thus, the discrepancy between the instruction and the pilot-side confirmation goes unnoticed and the pilot's action results in an incident.

As for background factors for errors that air traffic controllers make, 31 factors also affected pilots; namely those related to personal and interpersonal factors, mechanical system reliability and problems, and weather. Aside from these, 33 factors were found in 545 incidents (121 incidents in tower control, 142 incidents in approach control, and 282 incidents in en route control), although there are some differences among the types of air traffic control service (see Table 3).

Here, we will have a look at a few factors that I believe are particularly important in air traffic control service.

Concerning near misses (planes coming abnormally close to each other) and conflictions (abnormal intervals or spacing), there are no objective standards. A near miss is a serious situation in which the pilot fears that his plane may collide with or brush against another aircraft in flight. A confliction is a situation in which the proximity to the other aircraft is not as serious a concern as with a near miss, but in which the planes come within less than the minimum distance between the aircraft that should be maintained by the air traffic controller to avoid collisions (a distance referred to as the traffic control interval or separation). The distinction between these two situations, however, is not necessarily clear.

Also, even if the blips on the radar screen that represent two aircraft pass over each other (after the radar blips of two planes overlap on the radar screen, they can be distinguished by their directions), it does not necessarily indicate that a confliction, near miss, or collision has occurred. This is because the blips appear much larger than the actual size of the aircraft and, even though the two-dimensional distance between the two aircraft on the flat radar screen is less than the control distance, the planes are generally at different altitudes.

3.5 Mishearing in communications with pilots

All things considered, the important errors in incidents that involve human factors related to air traffic control communication occur during communication between pilots and air traffic controllers. Typical examples are unclear pronunciation or mishearing that often occur for the 'L' and 'R' sounds in the English words 'left' and 'right' or the words 'two' and 'three' or 'four' and 'five', which begin with the same consonant.

Each aircraft is assigned a call sign for communication with air traffic control units. For commercial airlines, the call sign is usually the three-letter code for the airline company plus the flight number of the plane. For civilian light aircraft, the call sign is usually the letters JA followed by a four-digit registration number. Military aircraft have call signs but they are assigned in an irregular way.

During communication between air traffic controllers and pilots, whatever the case, the call sign is given first.

When aircraft that have similar call signs are being handled at about the same time by the same controller, there is a high probability of the controller making errors of misspeaking as a result of jamming, mistaken impression, or preconception. When a plane that has a call sign similar to the one being instructed mistakenly follows the instructions intended for the other plane, confusion arises, such as a narrowing of the control distance between it and other aircraft.

An example is a case in which JAL flight 42 and JAL flight 427 were heading in the same direction from the same airport at the same time. These two call signs, expressed as 'four-two' and 'four-two-seven' in spoken form, are easily mistaken one for the other and caused confusion among the pilots of the two planes and the air traffic controller. (This incident actually occurred and was reported by a pilot.) While some means of dealing with such cases in the instructions is also needed, consideration should be given to the naming of flights by the airline companies so as to eliminate this problem.

There is also an incident in which two C-141 aircraft whose call signs differed by a single numeral (T50245 and T50246) were flying towards each other in the same air route. The air traffic controller, who had just started his work shift and was suffering from lack of sleep, spoke the wrong call sign in an instruction, with the result that the two planes were flying at the same altitude. The distance between them at the closest point was 2 nautical miles and collision was averted by giving

Table 3 Factors for quantification method III, air traffic controller related human-factor incidents (tower, terminal, en route)

	Tower
Total number of incidents:	121
Total number of events:	202

Personal factors
 Experience — Insufficient experience / unfamiliarity
 Physiological factors — Fatigue / lack of sleep
 Psychological factors — Inattentiveness (state of reaction after stress) / tedium

Team co-ordination — —

Meteorological factors — Fair / clear / good weather conditions
Adverse weather conditions / rain / snow
In clouds
Poor visibility / fog / haze
—
—

Control Equipment — —
—
—

Special circumstances — —
US Military jurisdiction airspace
Common-use military and civilian airport
—
Insufficient co-ordination
Shifts change / insufficient briefing
In training as controller
Insufficient / incomplete training and education as controller
Work environment
—
Language problem / foreign pilot
Similar call signs
Difference in awareness between controller and pilot
No radio, lost communication
—
—
Insufficient weather information
Aircraft flying head-on / priority runway
Climb rate / descent rate / speed / turning circle (miscalculation)
Emergency aircraft / non-standard flight / VIP flight
Special flight procedure / in inspection flight
—
Circling approach
SID (Standard Instrument Departure Route)
In visual approach
Cross runways
Request from pilot
Slow action by pilot

Terminal 142 225	Enroute 282 437
Insufficient experience / unfamiliarity	Insufficient experience / unfamiliarity
Fatigue / lack of sleep	Fatigue / lack of sleep
Inattentiveness (state of reaction after stress) / tedium	Inattentiveness (state of reaction after stress) / tedium
Excessive reliance / interdependency / excessive concern / misplaced deference	Excessive reliance / interdependency / excessive concern / misplaced deference
Fair / clear / good weather conditions	Fair / clear / good weather conditions
Adverse weather conditions / rain / snow	Adverse weather conditions / rain / snow
In clouds	In clouds
Poor visibility / fog / haze	—
Cumulonimbus / thunder clouds	Cumulonimbus / thunder clouds
Turbulence / wind shear	Turbulence / wind shear
No radar	No radar
No ARTS / no RDP / no FDP	No ARTS / no RDP / no FDP
Radio malfunction / radar out / RDP out / communication equipment malfunction	Radio malfunction / radar out / RDP out / communication equipment malfunction
—	Ocean control / via radio
US Military jurisdiction airspace	US Military jurisdiction airspace
Common-use military and civilian airport	—
Near boundary	Near boundary
Insufficient co-ordination	Insufficient co-ordination
Shifts change / insufficient briefing	Shifts change / insufficient briefing
In training as controller	In training as controller
Insufficient / incomplete training and education as controller	Insufficient / incomplete training and education as controller
Work environment	Work environment
Inconsistency regarding control language / manual incomplete	Inconsistency regarding control language / manual incomplete
Language problem / foreign pilot	Language problem / foreign pilot
Similar call signs	Similar call signs
Difference in awareness between controller and pilot	Difference in awareness between controller and pilot
No radio, lost communication	No radio, lost communication
In radar vector	In radar vector
At limit of radar coverage	At limit of radar coverage
Transponder out / no transponder	Transponder out / no transponder
Insufficient weather information	—
Aircraft flying head-on	Aircraft flying head-on
Climb rate / descent rate / speed / turning circle (miscalculation)	Climb rate / descent rate / speed / turning circle (miscalculation)
Emergency aircraft / non-standard flight / VIP flight	Emergency aircraft / non-standard flight / VIP flight
Special flight procedure / in inspection flight	Special flight procedure / in inspection flight
Direct route / irregular route	Direct route / irregular route
Circling approach	—
SID (Standard Instrument Departure Route)	SID (Standard Instrument Departure Route)
In visual approach	—
—	—
Request from pilot	Request from pilot
Slow action by pilot	Slow action by pilot

	Tower
Total number of incidents:	*121*
Total number of events:	*202*

	Pilot's unfamiliarity
	Violation of controller instructions by pilot
	Pilot in training / in route check flight
	Heavy workload
	Light workload
Performer	Controller (performer)
	Other controller (performer)
	Pilot
Information source	—
	—
	Instructions or communication from controller
	Communications by pilot
	Visual
	Other controller (information source) / land line
	Manual / NOTAM / weather information
Information receiving	Appropriate (information receiving)
	Did not see / hear
	Failure to see / hear / notice
	Saw incorrectly / heard incorrectly / mistaken impression
	Excessive focus
	Preconception
	Delay (information receiving)
	Could not see / could not see clearly / could not hear / could not hear clearly
Decision	Appropriate (decision)
	Forgot (decision)
	Easy going / assumption
	Wrong conviction / hasty conclusion / one's own interpretation / wrong expectation
	Vacillation (decision)
	Fixation (decision)
	Unreasonableness / high-handedness
	Delay (decision)
	Difficult (decision) / inability to decide
Action / instruction	Appropriate (action / instruction)
	Forgot (action / instruction)
	Instruction not received clearly / unclear pronunciation / aimlessness / sat idle
	Incorrect action or instruction / mistaken or inappropriate action or instruction / misspeaking / wrong expectation
	Fixation (action / instruction) / persistent action or instruction
	Recklessness / forcefulness / excessive action / over-tightening
	Delay (action / instruction) / insufficient action or instruction
	Difficult (action / instruction) / inability to act or instruct
Event critical phase	Information source (E)
	Information receiving (E)
	Decision (E)

Terminal	Enroute
142	*282*
225	*437*

—	—
Violation of controller instructions by pilot	Violation of controller instructions by pilot
Pilot in training / in route check flight	—
Heavy workload	Heavy workload
Light workload	Light workload
Controller (performer)	Controller (performer)
Other controller (performer)	Other controller (performer)
Pilot	Pilot
Radar screen	Radar screen
Strip	Strip
Instructions or communication from controller	Instructions or communication from controller
Communications by pilot	Communications by pilot
Visual	Visual
Other controller (information source) / land line	Other controller (information source) / land line
Manual / NOTAM / weather information	Manual / NOTAM / weather information
Appropriate (information receiving)	Appropriate (information receiving)
Did not see / hear	Did not see / hear
Failure to see / hear / notice	Failure to see / hear / notice
Saw incorrectly / heard incorrectly / mistaken impression	Saw incorrectly / heard incorrectly / mistaken impression
Excessive focus	Excessive focus
Preconception	Preconception
Delay (information receiving)	Delay (information receiving)
Could not see / could not see clearly / could not hear / could not hear clearly	Could not see / could not see clearly / could not hear / could not hear clearly
Appropriate (decision)	Appropriate (decision)
Forgot (decision)	Forgot (decision)
Easy going / assumption	Easy going / assumption
Wrong conviction / hasty conclusion / one's own interpretation / wrong expectation	Wrong conviction / hasty conclusion / one's own interpretation / wrong expectation
Vacillation (decision)	Vacillation (decision)
Fixation (decision)	Fixation (decision)
Unreasonableness / high-handedness	Unreasonableness / high-handedness
Delay (decision)	Delay (decision)
Difficult (decision) / inability to decide	Difficult (decision) / inability to decide
Appropriate (action / instruction)	Appropriate (action / instruction)
Forgot (action / instruction)	Forgot (action / instruction)
Instruction not received clearly / unclear pronunciation / aimlessness / sat idle	Instruction not received clearly / unclear pronunciation / aimlessness / sat idle
Incorrect action or instruction / mistaken or inappropriate action or instruction / misspeaking / wrong expectation	Incorrect action or instruction / mistaken or inappropriate action or instruction / misspeaking / wrong expectation
Fixation (action / instruction) / persistent action or instruction	Fixation (action / instruction) / persistent action or instruction
Recklessness / forcefulness / excessive action / over-tightening	Recklessness / forcefulness / excessive action / over-tightening
Delay (action / instruction) / insufficient action or instruction	Delay (action / instruction) / insufficient action or instruction
Difficult (action / instruction) / inability to act or instruct	Difficult (action / instruction) / inability to act or instruct
Information source (E)	Information source (E)
Information receiving (E)	Information receiving (E)
Decision (E)	Decision (E)

	Tower
Total number of incidents:	*121*
Total number of events:	*202*
	Instruction / action (E)
Incident critical phase	Information source (I)
	Information receiving (I)
	Decision (I)
	Instruction / action (I)
Reason for recovery	Controller himself (recovery)
	Other controller (recovery)
	Action by pilot
	—
	Good fortune
Degree of danger	Virtually none
	Fairly little
	Considerable
	Serious

instructions for both planes to make turns to the right. The degree of danger for this incident was 'serious'.

In another incident, the air traffic controller had issued clearance for take-off to a VFR light aircraft that was second in line (No. 2 plane) but had a call sign similar to the plane that was first in line (No. 1 plane). The mistake was noticed when the pilot confirmed the instruction and the clearance was immediately cancelled and clearance was issued for the No. 1 plane. The degree of danger in this case was 'virtually none'.

Through confirmation by read-back, the air traffic controller can correctly recover from mishearing by the pilot, and the pilot can correctly recover from misspeaking by the air traffic controller. If both controller and pilot fall into 'wishful hearing', however, then the danger of near misses or confliction is high.

Concerning matters for which a person has a strong expectation or prediction, the person may tend to hear what is expected or predicted, even if what is actually said differs somewhat from what is expected or predicted. In other cases, although part of the information is actually lost, the listener receives the mistaken impression of having understood, based on the contextual relationship of the elements that are heard.

For example, clearance for take-off is usually issued after information on wind direction and wind speed, so it sometimes happens that a pilot takes it that take-off clearance has been given even though actually only the wind direction and speed information was reported, or that information that differs from the information that is normally reported is taken as that normally reported information. For example, during a landing, if the clearance to descend to a particular altitude at a certain point is given, the pilot may assume that he has heard the clearance to

Terminal	Enroute
142	282
225	437
Instruction / action (E)	Instruction / action (E)
Information source (I)	Information source (I)
Information receiving (I)	Information receiving (I)
Decision (I)	Decision (I)
Instruction / action (I)	Instruction / action (I)
Controller himself (recovery)	Controller himself (recovery)
Other controller (recovery)	Other controller (recovery)
Action by pilot	Action by pilot
Warning device	Warning device
Good fortune	Good fortune
Virtually none	Virtually none
Fairly little	Fairly little
Considerable	Considerable
Serious	Serious

descend to the usual altitude, even in the unusual cases where clearance was given to descend to a different altitude.

While this type of phenomenon is often experienced in our daily lives, 'wishful hearing' is especially likely to occur in air traffic control situations, where communication is very formal and employs a limited set of terms.

While a VFR plane in training (plane A) was flying the left downwind leg (the part of the flight path that is parallel to the runway and is used when turning left to begin the approach to the runway – see Fig. 21) to runway 32 and an IFR plane (plane B) was making an ILS approach, another VFR plane (plane C) requested to proceed directly to the left base leg directly, without flying the left downwind leg. The 'base leg' is the part of the flight path before the final that is perpendicular to the runway and from which the final approach is entered by making a turn, in this case to the left. The 'final' is the flight path that is aligned with the runway and is the final stage of the landing.

The tower local controller informed plane C that the request could not be granted because plane A was already in the left base leg and instructed plane C to enter the right base leg. However, the pilot of plane C misheard 'right' as 'left', and read back the instruction as 'Roger left base'. The controller, who was in training, failed to notice this incorrect action.

On the other hand, the controller informed plane A, which was on the left downwind leg, that plane B was making an ILS approach and instructed plane A to 'Make short approach' (make a tighter turn than usual to begin the approach). However, plane A was a training plane, and it entered the standard left base leg rather than making the short approach. (For each runway of an airport, standard flight paths called

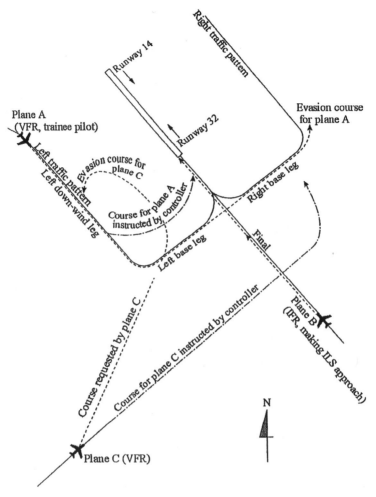

Fig. 21 Flight paths for near-miss incident that involved 'wishful hearing' by both air traffic controller and pilot

flight patterns are defined, including the downwind base leg, final approach, and so on.) Because plane C entered the left base leg against the intentions of the controller, it came close to plane A. By the controller's lastditch measures, plane A was able to evade closing with plane C, but then it had no separation with plane B, making an ILS approach, then entered the final approach. The controller instructed plane A to cross over the final approach path.

This incident, for which the degree of danger was 'serious', is a typical example of simultaneous 'wishful hearing' on the part of both the controller and the pilot of plane C. The time before this situation occurred was the peak (busiest) period for the controller, and he was in a state of post-stress carelessness. When plane A entered the base leg instead of making the short approach, he sat idle without taking the proper steps.

The critical phase that determined the incident was the information receiving phase for the trainee controller, when he misheard the read-back from the pilot of plane C owing to his preconception. The background factors included the post-stress carelessness of the controller, that the controller was in training, and that the pilot of plane A was in training.

3.6 Danger created by inappropriate terminology

The terminology and control procedure for air traffic control are set by international standards. If non-standard terminology is used when providing instructions or traffic information, the controller's intended meaning may not be correctly conveyed to the pilot.

In one reported incident, the term 'maintain' was used when reporting the altitude of a related aircraft, with the expression 'Traffic one o'clock 15 miles (aircraft type) maintaining flight level 310'. This means that the related aircraft is in the one o'clock direction ($30°$ to the right), 15 ground miles away, and flying at an altitude of 31 000 ft. Because 'maintain' should be used only when giving clearance, the pilot had a mistaken impression of the information as an altitude clearance for his own plane and so climbed.

The controller himself noted the problem and quickly gave instructions for the plane to return to its original altitude, but the degree of danger was still 'considerable'.

There are also three other incidents in which similar inappropriate use of the term 'maintain' was an important background factor, and three of the total four incidents occurred in the same office.

The single reporter of those incidents pointed out that 'Rather than differences in the interpretation of the air traffic control terminology in high dimensions, a difference in the degree of understanding and interpretation can be seen for each person or office or between controllers and pilots, and I believe that this is a major element of danger'.

In another incident, a controller intended to instruct a pilot immediately to descend so as to stay in a proper relationship with other aircraft, but he did not add the word 'immediately' to his instruction. Because of that, the pilot understood it to mean that it was all right to maintain his current altitude for a while before beginning the descent, with the result that the separation was lost.

3.7 Discrepancies in recognition between air traffic controllers and pilots

In addition to inappropriate use of air traffic control terminology, discrepancies in recognition between air traffic controllers and pilots concerning control and aircraft operation is another background factor to the occurrence of unsafe events.

An L-1011 commercial airline was making an approach to an airport that did not have an approach control area, and so was flying towards the approach point from a point approximately 35 nautical miles from the airport and descending. Directly opposite the civilian plane, about 25 nautical miles on the other side of the approach point, was a military P-2J plane flying under visual flight rules.

A request to go to instrument flight rules was received from the military plane, and clearance to do so was given. At the same time, the P-2J was instructed to fly to the destination at the high altitude of 9000 ft. The controller in the en route radar position, having already instructed the civilian plane to descend to 6000 ft for its approach to the airport, thus decided that, when the two planes came close to each other at the approach point, the civilian plane was probably descending to below 8000 ft.

However, the civilian plane had, without making a special report of the fact, maintained the altitude of 10 000 ft instead of descending, and was nearing the approach point at that altitude. Soon, after seeing the situation, the civilian airliner suddenly began a descent, and, although there was an altitude separation between it and the military plane, the radar distance (i.e. the horizontal separation of the two aircraft as displayed on the radar screen) became zero.

When, after the landing, the pilot of the airliner was asked in a phone call why he had not descended from 10 000 ft, he revealed that the pilots of turbojet planes were bound by the internal airline company operation manual to fly level for a time at an altitude of about 10 000 ft in order to reduce to the speed limit specified by air law, even when flying outside the approach control area.

The airspace for which a speed limit is stipulated by the Civil Aeronautics Law (Article 82-2) is the airspace below 10 000 ft within the approach control area and the speed limit is 250 knots (about 276 mile/h) indicated airspeed (the flying speed relative to the air around the plane).

Although this air traffic controller, too, knew that the speed limit is specified by the air law within the approach control area, the airport at which the civilian plane was landing in this incident did not have an established approach control area. The controller believed that the civilian plane would immediately descend to 6000 ft as instructed, because he 'did not know the internal operation manual of each airline'.

Civilian airlines have this kind of internal operation manual because it is difficult for turbojet aircraft both to reduce speed and to reduce altitude at the same time. However, as this controller said, air traffic controllers are not necessarily aware of these kinds of manual. In cases such as this, when an aircraft does not intend promptly to follow instructions, the fact should be reported to the air traffic controller, as one would expect. Also, the obligation to make such a report should be added to the internal company operation manual.

There is also an incident in which interdependency and excessive reliance stemming from the differences in understanding of three parties, the air traffic controller, a VFR plane, and an IFR plane, resulted in the two planes crossing paths with 500 ft difference in altitude between them. The degree of danger for this incident was 'fairly little'.

On a day of good weather conditions, a VFR plane requested radar monitoring by the terminal radar controller, and, when the work was being done at 15 000 ft, an IFR plane came near. The controller provided the IFR plane with the appropriate traffic information, and, because the VFR plane reported that he saw this, properly thought that he would take evasive action. The pilot of the VFR plane, however, had it in his mind that, because he was on radar monitoring, if evasive action were necessary, the air traffic controller would issue instructions for that.

After that, there was only a report by the IFR plane that it had come quite close to the VFR plane.

Pilots, including the pilots of commercial airlines, generally recognize that, when an IFR plane and a VFR light plane come close to each other, evasion is entirely and in every case the responsibility of the VFR plane. None of the stipulations of the air law, however, specify right of way with respect to planes flying under different flight rules (see Article 71-2 and Article 83 in the Civil Aeronautics Law of Japan and Article 180 and those that follow in the Civil Aeronautics Regulations).

Nevertheless, large aircraft, considering their manoeuvring capabilities, often have difficulty in manoeuvring to avoid other aircraft that have approached to within a certain distance. Although large aircraft also can slow down relatively quickly by using speed brakes, it is still easier for light planes to take evasive action. Because of that fact, the perception of the danger of close encounters between IFR planes and VFR planes is quite different for the two types of aircraft.

In an incident in which an IFR plane and a VFR plane crossed paths at a distance of 500 ft at the same altitude (8500 ft), the feeling in the IFR plane was that 'It was dangerous, because the planes came quite close to each other', but for the VFR plane, 'I didn't feel danger because I believed complete evasion was possible'.

3.8 Problems near the boundaries between air traffic controlled airspaces

When an aircraft that is flying in the controlled airspace for which the air traffic controller is responsible passes into the controlled airspace of another controller, the first controller transfers the control of that aircraft to the second controller.

The transfer of control, in principle, takes place at the boundary between the two controlled airspaces. When a plane that is flying in the controlled airspace of a controller is going to fly into another controlled airspace, the first controller performs a radar hand-off before the plane reaches the controlled airspace boundary. In transferring control, the controller reports to the controller of the other airspace (i.e. the inheriting controller) the distance and direction from the radar fix of the aircraft (the specific point indicated on the radar scope used for radar identification and radar transfer, which is controlled electrically or mechanically), magnetic course, and other such information. He also transfers the radar ID, an action referred to as radar hand-off.

The controller then instructs the pilot of the plane immediately to switch the communication radio frequency to the frequency of the inheriting controller. Thus, after the frequency switching, the controller can no longer communicate directly with the plane, even though it is still flying within the airspace for which that controller is responsible.

Accordingly, the transfer of control must be done at a time when the separation between the plane and other aircraft is being maintained. In actual practice, however, it sometimes happens that, for some reason or other, it becomes necessary to give instructions to the plane after the transfer of control.

In such cases, the only thing that can be done is for the controller that has relinquished control to notify the inheriting controller of the fact so that the inheriting controller can then give the instruction. However, communication traffic is often congested and it is not always possible to establish communication with the inheriting controller immediately.

Even assuming that immediate communication is possible, it is still unavoidable that this indirect communication between the relinquishing controller and the aircraft will be slower than direct communication. During that necessary delay, the positional relationships between the aircraft and others are becoming more and more worrisome.

Because of this situation, there is a high possibility of incidents occurring in the vicinity of the boundary between controlled airspaces. This problem at the time of control transfer can be raised as an example of a human-factor incident involving air traffic controllers.

One such case is the following near-miss incident that occurred near the boundary between Narita airspace and Haneda airspace and involved a plane that was flying in Narita airspace under control by Haneda air traffic control and an aircraft under Narita control. The degree of danger for this incident was 'serious'.

Because there was a request from the Haneda air traffic controller to have plane A, which was arriving at Haneda Airport, fly into the Narita control airspace so as to avoid adverse weather conditions, permission was issued immediately. On that day, a large cumulonimbus had formed in the arrival route to Haneda, and other aircraft had also been instructed to fly a route for evading the adverse weather conditions.

Because plane A was receiving instructions on the radio frequency of the Haneda control even after entering Narita airspace, I gave issued clearance via the Haneda controller for plane A to descend to 10 000 ft. On the other hand, I instructed plane B, which was flying in the same route as plane A in the opposite direction for arrival at Narita, to descend to 11 000 ft.

Because at that point plane A was at a lower altitude than plane B, it was natural to expect that, by about the time plane B was reaching 11 000 ft, plane A was reaching 10 000 ft.

As it actually happened, however, plane A's descent was very delayed, with the result that the two planes were flying head-on at about the same altitude. Because I could not issue instructions to plane A directly, the Haneda controller and I together took measures to avert a collision, but these adjustments took some time.

Also, if the controllers are in the same control room, one controller can shout 'Bank right!' or 'Stay at 11 000 ft!' and the instruction can be issued immediately. With this kind of co-ordination, however, when the controllers are in different places, a call must be made with a special telephone before the co-ordination can take place, and that takes time.

Just before the two planes came close to each other, the Haneda controller gave evasion instructions to plane A and I gave instruction to plane B, but the instructions were, in effect, late and the radar targets of the two planes completely overlapped at about the same altitude. Everyone in the radar room was holding his breath with eyes riveted on the radar screen. Immediately after that, the merged radar targets slowly separated into two again, leaving everyone weak in the knees. We had been saved.

The background factors here were the cumulonimbus, the pilot's request, and wrong expectation.

3.9 Oceanic control service via radio station

The range of air route surveillance radar is a radius of approximately 200 nautical miles, so at this time the work of oceanic control service is performed on the basis of information obtained from the operations table and communication with the aircraft alone. That is to say, radar is not available as a visual source of information, and there is only the (flight) strip. Moreover, communication with distant aircraft is accomplished indirectly by shortwave radio by way of the air traffic control communications specialist. Controllers call this 'via radio'.

Thus, even in cases where the controller wants to give urgent instructions to an aircraft, it must be done through the air traffic control communication specialist, and if, at that time, the communication specialist is busy with another communication, the instruction will be delayed. Also, because high-frequency radio transmission is not as clear as superhigh-frequency radio transmission, mishearing can easily occur.

Furthermore, as the other units with which the communication co-ordination must be conducted may be the air traffic control units of a different country, there is ample opportunity for language problems or differences in awareness to occur.

In one reported incident, two planes that had had separation lost the separation just after a shift change of the oceanic controller. Although one of the planes was instructed to climb so as to avoid collision, communication was not possible for a short time. The degree of danger for this incident was 'considerable'.

In another incident, after a flight of three F-101 jet fighters was instructed to climb to 33 000 ft, it was noticed that a flight of four F-86s was flying head-on towards the flight of F-101s at an altitude of

31 000 ft. The air traffic controller wanted immediately to cancel the instruction, but it took some time because the instruction had to be relayed through the radio station. By the time the cancellation of the climb instruction had reached the flight of F-101s, they had already begun the climb, so they could only disregard the cancellation: 'All that could be done was to hope that the flight reached 33 000 ft, and both hands were in a cold sweat'. The degree of danger for this incident was 'serious' and the reason for recovery was 'good fortune'.

3.10 Highly dangerous incidents due to cumulonimbus

Even with today's highly developed aviation technology, weather conditions are a factor that greatly affects the operation of aircraft, as was explained in Chapter 2. Accordingly, weather conditions also greatly affect the air traffic control service. In particular, cumulonimbus, thunderclouds, and turbulence interfere with flight safety and comfort, so pilots request changes of courses that have been assigned in advance by air traffic controllers and request radar guidance. This not only increases the workload of air traffic controllers but also serves as a background factor that leads to delayed instructions and acts of forgetting.

In such cases, the air traffic controller must instruct the plane to take a different altitude or a different route than they ordinarily would if the conditions presented no obstacles. Therefore, in the same way as pilots in the case of unexpected changes in runways or routes, the controller must change the plan that he had already formed in his mind.

Particularly when cumulonimbi form in an airspace, there is a remarkable increase in the workload of air traffic controllers, because, one after another, many aircraft request changes in route or altitude so as to avoid the clouds. Furthermore, the radar echo of cumulonimbus is not so clear on the controller's radar screen as it is on the on-board radar screen of the aircraft, so it is difficult to judge the best course for evading the clouds.

The following are examples of this type of incident.

There was a request from a pilot for a route change because of cumulonimbus. The originally planned ILS circling approach was abandoned and the plane was guided to the downwind leg for a visual approach. An 'ILS circling approach' is one in which, when the ILS is installed only in one direction, the approach is made on the ILS course, but, when the runway becomes visible, a turn is made and a visual landing is made on a different runway, which is usually the runway in the opposite direction from the one on which the ILS is installed. Concentrating on that job, however, the controller forgot to co-ordinate with the local controller (i.e. the tower controller), and, because of that, the local controller gave clearance to take off to a departing plane that may have come close to the landing plane.

Although the separation between the two planes ran short, the degree of danger was 'fairly little', because the weather presented visual meteorological conditions and the two aircraft were able to see each other.

The next example occurred near the boundary between the controlled airspaces of the arrival controller of the terminal control and the area control centre (the unit that performs air route control). The degree of danger for this incident was 'serious' and the reason for recovery was 'good fortune'. The 'good fortune' reason for recovery in the field of air traffic control service refers to incidents that narrowly escape development into accidents without averting measures by the air traffic controller himself, other controllers, or the pilot, and without the functioning of a warning device.

Receiving a request from plane A, which was approaching for a landing, for a heading because the normal guided course could not be maintained owing to the need to avoid cumulonimbus, the arrival controller gave approval. The controller in the arrival control obtained approval from the departure controller in the same terminal radar control room and had the plane descend for the approach in that airspace.

In the departure control airspace, there was plane B, which was climbing and flying towards plane A in the opposite direction. The departure controller, however, failed to notice that the two planes were coming close, because his radar screen was, as always, set to a narrow range and plane A was not on his screen. He thus forgot to establish the separation between plane A and plane B before handing off control of plane B to the area control centre.

The arrival controller who was handling the control for plane A, without knowing that control for plane B had been handed off to the area control centre, requested the departure controller temporarily to stop the climbing of plane B because of the proximity of plane A and plane B. At that time, however, the co-ordination after transfer of control took some time and the instructions were not in time.

Although plane A was informed of the departing plane, the two planes could not see each other because they were both in clouds and continued their descent and climb, and narrowly escaped danger by crossing paths with a mere 300 ft difference in altitude. The background factors were cumulonimbus, pilot request, nearness to the control airspace boundary, flying in clouds, and insufficient communication/co-ordination.

Another type of incident to consider is one in which strong thunderclouds in the vicinity of an airport affect the radio signal of the ILS approach system. Because of that, a plane made its approach with an incorrect approach course and approach angle, descending excessively and coming close to an obstacle on the ground. In some cases, there is danger of such incidents developing into major accidents.

3.11 Danger due to poor visibility within clouds

Like cumulonimbus, flying within clouds is a background factor that cannot be overlooked, particularly in the context of the investigations in this research, which was conducted when it was not obligatory for commercial aircraft to be equipped with TCAS.

If the weather is fair and clear (visual meteorological conditions), even aircraft that are flying on instruments are obligated to maintain a visual lookout for the situation outside the aircraft (Civil Aeronautics Law of Japan, Article 71-2). Therefore, even if, for example, a confliction or a near miss occurs owing to inappropriate instructions by the air traffic controller, evasive action by the pilot can be expected.

When flying within clouds, however, the pilot cannot see outside the plane, so it is not possible for the pilot to be aware of the danger from other nearby aircraft unless the plane is equipped with a 'Traffic Alert and Collision Avoidance System' (TCAS),[1] and evasive action by the pilot cannot be expected. Accordingly, the inappropriate instructions by the air traffic controller are the determining factor in the incident.

In such cases, the radar targets for the two aircraft overlap on the radar screen, even if there is an altitude difference between the planes. Therefore, the controller is in a sweat and, not until the radar targets separate again, does he release a deep breath of relief, but the cockpit crews in the planes most often do not even notice the situation. A near miss that goes unnoticed by pilots flying within clouds can probably be said to have the highest degree of danger. The example presented earlier in the section on cumulonimbus in which the degree of danger was 'serious' is one such case.

There are also many reported incidents involving aircraft coming close to each other while flying in clouds without the crews being aware of it because they could not see the other planes and without any report issued, resulting in a 'serious' degree of danger and 'good fortune' as the reason for recovery.

In one such incident, three planes that were under guidance by the terminal radar controller because of cloud cover down to about 2000 ft were at the same radar distance of 1 nautical mile. The background factors included, in addition to flying in clouds, insufficient ability of the air traffic controller. The reason for recovery was 'good fortune'.

The following is another reported incident.

An IFR commercial airliner was being guided to the final approach course and a departing freighter plane was climbing under a Standard Instrument Departure (SID). The approach controller (one of the controllers performing terminal radar control) handed off the approaching plane off to the arrival controller (one of the controllers performing terminal radar control, also called the feeder controller) without establishing the appropriate separation between the two aircraft and without informing the arrival controller of the departing plane.

When the arrival controller noticed the radar target and asked the approach controller, the altitude difference of the two planes was from 100 to 120 ft, and there was no time to issue instructions for evasive measures. Here, too, the reason for recovery was 'good fortune'.

In addition to that case, there are other incidents, such as when the radar targets of two aircraft, a DC-8 international airliner and an L-1011 domestic plane that took off behind it, nearly overlapped completely at 20 000 ft because an air traffic controller, who had received his qualifications for the terminal radar control just before the case, had a wrong expectation concerning the climb rate of the first plane and was slow in taking evasive measures. In that incident, although the traffic information was communicated to the two planes, both planes were in clouds and could not see each other. The main background factors were the wrong expectation regarding the climb rate of the first plane, insufficient experience, and flying in clouds. The reason for recovery was advice from another controller.

3.12 Pitfalls of good weather

When the weather is good (i.e. visual meteorological conditions exist), danger can be averted or reduced by evasive action on the part of the pilot. On the other side of the coin, however, the workload of the air traffic controller is increased because there are many flights for training, aerial photography, and other small aircraft (special flights), and, because small planes fly in unpredictable ways, the latent danger is higher.

During the work shift change meeting, I had heard that there was only a single radar target for a plane on the radar screen. That plane was being vectored to the final approach course, and, 2 min later, radar targets for two planes appeared from the middle of the radar screen at the same time, moving in opposite directions. At that time, the radar antenna had not been updated and the reliability of the terminal radar equipment was poor, with ghost targets often appearing. Because of that, concerning one of them, the departure information for VFR planes was obtained from the tower, but the other target was taken to be a ghost.

I was vectoring the one plane that was handed off to me, and, about the time it was beginning the final approach, it came close to what I thought was a radar ghost. I hurriedly asked the tower controller and the co-ordinating controller and learned that clearance for approach had been given by another controller before the shift change. I issued instructions for evasion to the plane that was being guided. However, the weather was clear and the pilot saw what was happening. He had already taken evasive action.

The degree of danger in that incident was 'considerable'. The aircraft for which the target appeared later on the radar screen was flying in the airspace directly above the radar antenna at the time of controller work shift change, and so, because of the way the radar works, the target for that plane was not shown on the screen for a short time.

Failure to pass on information at the work shift change and the low feeling of trust in the radar equipment because of repeated appearances of ghost blips are the factors working in the background of this incident.

Aerial photography is done in good weather. Sometimes, however, it is not possible to follow the flight plans submitted in advance because of the circumstances or how objects can be viewed from the air.

Although with the original flight plan there was a sufficient separation between departing planes, the flight plan was changed without informing the air traffic controller. As a result, while the controller's attention was briefly on another plane, the plane crossed paths at the same altitude as a departing plane at 8500 ft with a separation of 0.5 nautical miles. The two planes saw each other in advance and the degree of danger was 'fairly little' in this incident.

The following is an example in which 'touch and go' training (a landing practice exercise in which the plane touches down on the runway and then immediately goes on to take off again) was being conducted because of good weather.

There were two intersecting runways (cross-runways) under the priority runway procedure. On the south leg of the traffic pattern for one of those runways, runway 12, were three VFR training aircraft that were engaged in 'touch and go' practice. On the final approach course of the other runway, runway 27, were two VFR training aircraft that were doing 'touch and go' practice. The air traffic controller believed that, because of the good weather, each of the training planes could understand the situations of the other training planes, and, because the workload was high and there was little spare time, clearance for a touch and go was given to the first plane in the landing order of those that were in the final approach of runway 27, but without providing traffic information. However, the plane that was on the final approach passed under a plane that was flying the base leg of runway 12.

The degree of danger was 'serious'. The controller's wrong conviction during the decision phase was the determining factor in this incident.

3.13 Wrong expectation by air traffic controllers

In performing air traffic control, controllers, to some extent, make predictions concerning the climb rate and descent rate, speed, and turning circle of each aircraft when maintaining an appropriate separation between the aircraft. When the climb rate or turning circle of the aircraft under control differs from what the controller takes into mind, however, the degree of danger becomes high when the controller is concentrating on another matter.

The climb rate of an aircraft is affected by the type of plane, weather conditions (particularly air temperature), the operating weight (the basic weight of the plane, including crew, plus the weight of the fuel being carried and the weight of the passengers and freight), and the speed and altitude of the plane. Generally, turboprop planes (airplanes that have turbine engines that drive propellers; the YS-11 is a plane of this type) have a low climb rate and jets have a high climb rate at low altitudes. Because the aerodynamic lift decreases as air temperature increases, the climb rate is usually poor in summer and relatively good in the winter months. If anti-icing equipment is activated, however, the climb rate is decreased.

Also, even for the same type of aircraft and the same weather conditions, a higher operating weight will result in a lower climb rate and a lower operating weight will mean a higher climb rate. Furthermore, from the relationship between the altitude at which the plane is flying and the climb rate, the climb rate is generally lower above the altitude of 25 000 ft.

At airports where international airliners and domestic airliners take off and land, in particular, the flight characteristics of each plane vary greatly with the destination, the take-off weight, the type of aircraft, and the airline company, so it is quite difficult for the air traffic controller accurately to predict the climb rate and other such characteristics of those respective planes. The turning circle is governed by the speed and banking angle of the aircraft, and differs with the way in which the pilot operates the aircraft, so there are also often discrepancies between the controller's predictions of this characteristic and the actual characteristic. Aircraft speed, as well, is greatly affected by the type of aircraft and high-altitude winds, and so is a background factor in wrong expectations.

One air traffic controller usually has control over many aircraft and cannot continuously concentrate attention on a particular plane alone. Climb rate and other characteristics that depend on aircraft type are knowledge that can be learned and must be learned. However, the fact that 60 of the 545 reported incidents have wrong expectations regarding these characteristics as a background factor shows that it is quite difficult for air traffic controllers accurately to predict the individually different flight characteristics of multiple aircraft under specific circumstances.

There is an incident in which a C-1 transport plane, which had taken off after a Boeing 737, cut across the path just in front of the Boeing 737 at the same altitude, because the tower controller had a wrong expectation about the climb rate and turning circle of the C-1. The visibility was good and the pilots of both planes were able to see the other plane, so the degree of danger was no greater than 'considerable'. If they had been flying in cloud, however, the two planes would not have been able to see each other even if traffic information had been given, so the degree of danger would have been higher. There are actual terminal radar control incidents in which the radar targets for two planes nearly overlapped on the radar screen and the degree of danger was 'serious'.

The following incident was also reported.

The pilot of plane A, a non-Japanese aircraft that was flying at 35 000 ft, made a request to the air route radar controller for permission to change altitude to 39 000 ft. A co-ordinator informed the radar controller that the request was approved, and, after confirming that plane B (a Japanese aircraft) passed by at 37 000 ft, the radar controller issued clearance to plane A for 39 000 ft and gave traffic information to the two planes.

After that, when the controller was concentrating on another plane, a warning device activated and he noticed that the two planes had a confliction. That happened because the climb rate of plane B had decreased suddenly and the climb rate of plane A had increased. Because the two planes could see each other, however, the degree of danger was 'fairly little'.

The Radar Data Processing system (RDP – a system for processing air route radar data) used in air traffic control is equipped with a warning device, but the Automated Radar Terminal System (ARTS – a terminal radar data processing system) has no warning devices. The fact that in this example the situation was brought to the air traffic controller's attention by the operation of a warning device points to the desirability of developing and installing a warning device for ARTS as well.

3.14 Slow action by pilots

The problems of climb rate, turning circle, and so on are not the only causes of wrong expectations on the part of the air traffic controller. Slow action by pilots is also a factor in the controller's problems. The biggest problem that faces the controller as the result of slow response of aircraft to the control instructions is in tower control.

Usually, departing planes begin their take-off as soon as they receive clearance to do so. Therefore, unless the air traffic controller specifically receives a report from a plane that has been cleared for take-off, he assumes that the plane has taken off promptly. Even if the take-off is a little delayed, thinking that it will take off in the next moment, the controller can only wait with increasing concern for the separation between the plane that is taking off and the plane that is landing. Thus, it is difficult to decide whether or not to have the landing plane execute a go-around.

If an arriving plane lands while a departing plane on the runway has not begun its take-off, the danger of collision arises, and, if the arriving plane executes a go-around when the departing plane is beginning its take-off climb, the two planes will come close to each other in the air. For that reason, the controller becomes unsure of what instruction to issue and at what time to issue it.

The pilot of the departing plane first increases the engine output to a certain extent to check the state of the engine acceleration, and then increases the output to the take-off level. If an abnormality appears in the acceleration stage, however, it takes a short time to decide what should be done, and the beginning of the take-off will be delayed, contrary to what the air traffic controller expects. In such cases, the workload on the pilot's side also increases, so at times the pilot may forget or delay the report to the air traffic controller or fail to hear communication from the air traffic controller.

Arriving planes, on the other hand, do not immediately begin the go-around operation when the separation with the plane in front is somewhat reduced because it would bring about delay beyond the planned arrival time and the concern of the passengers.

Go-around is one operation that is generally not desired. Once clearance to land is obtained, it is understood that it is all right to land and that there are no hindrances to landing, so this inclination is strong. Go-around is normally decided because of imminent danger. Unless there is impending danger, executing a go-around on the pilot's decision is all the more an invitation of danger.

If, in this way, the take-off of a departing plane is delayed without notifying the controller, even for just a slight amount of time, the separation will be narrowed and a dangerous situation may easily result. Because the separation between aircraft is relatively small in the vicinity of an airport, it is extremely important for maintaining safety that pilots immediately report any intention not to follow an instruction from air traffic control promptly. It is necessary that pilots fully understand this importance.

It is also important for air traffic controllers to give appropriate traffic information to the pilots of departing planes, like 'Take off immediately', 'There is an arriving plane XX miles from the runway', or 'If you cannot take off immediately, hold your position'.

'The movement of the departing plane was slower than expected, and when a fast-moving arriving plane landed, the departing plane was still moving on the runway'. That is a simple description, but the anxiety of the controller over the slow movement of the departing plane is evident. The degree of danger for this incident was 'serious'.

3.15 Danger of special flights by small aircraft

There are some out-of-the-ordinary types of flight, unrelated to the distinction between instrument flight rules and visual flight rules. Examples include aerial photography by the press and companies that use aircraft, surveying, power line patrols, crop dusting, search and rescue, training, and inspection flights by the flying corps of police organizations, fire departments, and the Maritime Safety Agency, as well as Air Self-Defence Forces scrambles (an emergency dispatch of aircraft that is executed when a plane of unidentified nationality violates the air defence identification zone) and air transportation of VIPs.

Except for the transportation of VIPs and Air Self-Defence Forces scrambles, most of these special flights are usually made by light aircraft or helicopters and take place within a control area, a control zone, or a positive control area. They thus have a large effect on commercial airlines.

The light aircraft used in these special flights, because of their purpose, change attitude instantly and fly complex patterns (repeatedly pass over a certain place on the ground, etc.). Thus, unless the air traffic controller pays constant attention to these flights, it is difficult to maintain separation with other aircraft. Also, there is frequent and lengthy communication with these aircraft because they frequently change route, altitude, and attitude.

In recent years, the use of multipurpose aircraft has increased and the number of special flights has been increasing. The resulting increased workload of the air traffic controllers has created a burden of mental stress for controllers.

I will present a few examples that show how difficult it is for controllers to predict the movements of aircraft that are engaged in special flights.

At the time when a VFR plane that was doing aerial photography at an altitude of 15 000 ft changed the communication radio frequency to that of the area control centre, an IFR plane was approaching. The radio frequency of the IFR plane was changed from the area control centre to the frequency of the terminal radar controller in the hand-off at an altitude of 16 000 ft.

Just as the heading of the approaching plane was changed while watching the direction in which the VFR plane was flying, the VFR plane suddenly increased altitude in the direction of the approaching plane so that the two planes were drawing near each other. Because the communication switchover had just been executed, however, the controller could not contact the plane, and, although the approaching plane was given traffic information and instructions to change heading so as to avoid collision, a confliction incident occurred. The degree of danger here was 'fairly little'.

There are also incidents, such as the following, for which the degree of danger was 'considerable'.

The tower controller approved a request from a VFR plane that was doing aerial photography in the vicinity of the airport to cross the final approach course. However, the VFR plane did not cross the final approach course immediately, so the controller instructed the VFR plane to stand by, because he felt there was danger of a close encounter with an approaching DC-9 commercial airliner.

Another near-miss incident experienced by a radar controller of an area control centre that at that time did not have an RDP involved a flight of three US military aircraft that were climbing to an altitude of 29 000 ft and a commercial airliner that was flying head-on towards them at 28 000 ft.

This incident occurred because the controller mistakenly believed that, when the leader of the flight of military planes acknowledged reaching the altitude of 29 000 ft, the other two planes in the formation flight had also reached that altitude. The controller was unaware that there is an altitude difference between the flight leader and the two accompanying planes during a climb. When the flight leader plane reached altitude, the two accompanying planes were still climbing, and the commercial airliner had to take evasive action, seeing that the military planes were cutting across its path in front of it.

The degree of danger in that incident was 'serious'.

3.16 Incidents that involve consecutive near misses

It sometimes happens that the action taken by an aircraft according to instruction given to prevent two planes from coming too near each other can affect the distance between that plane and a third plane, resulting in consecutive near misses. Such cases are referred to as consecutive near misses or consecutive conflictions and are generally associated with a high degree of danger. A typical example of consecutive incidents is the one shown in Fig. 21 on page 152, but I will describe other examples here.

Of the three planes involved, one, plane A, was a US military aircraft on a special flight (training). On the (flight) strip that was submitted for that plane before the flight, it was written that the plane would fly east and then, at a DME distance of 40 nautical miles, switch directions and fly west. 'DME distance' refers to direction measurement equipment, which determines the distance of the aircraft from a station on the ground by the response radio signal from the station to a query signal from the aircraft and displays the distance in the cockpit.

On the (flight) strip, eastward travel is indicated by a yellow colour and westward travel is indicated by a green colour. Course changes are indicated by letter descriptions, but the radar controller was late to notice that the plane would fly west during the flight owing to a preconception, because the (flight) strip was yellow.

Plane B, a Boeing 747 of a foreign airline company that was flying at the same altitude as plane A, 33 000 ft, was heading east, so the controller decided to instruct plane A to descend to 31 000 ft. However, the distance between the two planes was reduced to 30–40 nautical miles, because the controller could not get through to plane A on the radio telephone. Because of that, plane B was instructed to head due north, but plane B could not understand the reason for that and so the communication took some time.

Meanwhile, communication with plane A over the radio telephone became possible, so that plane was instructed to descend to 31 000 ft and turn left, and at that time plane A finally changed heading, but the distance between the two planes had closed to about 10 nautical miles. Because aircraft travel at a high speed in the air route, the planes continued in their courses for 5–10 nautical miles until the instructions were issued and the headings were changed. Because the air traffic controller was in training, the co-ordinator quickly instructed the radar controller that plane B should climb to 35 000 ft.

Although through that action it was possible to maintain the distance between plane A and plane B, the situation of plane B and plane C, a US military aircraft that was flying in the opposite direction at the same altitude, became severe. The degree of danger was 'serious'.

3.17 Survey results that overthrow current beliefs

We classified the 545 reported incidents that are related to air traffic control service into three groups according to the type of air traffic control under which they occurred (i.e. the incident critical phase), tower control, terminal radar control, and en route control, and compared and examined them from an integrated viewpoint. The first thing that can be said from the results is that, looking at the three types of control service comprehensively, the incidents associated with the 'serious' and 'considerable' degrees of danger make up some 70 per cent of the total.

The evaluation of the degree of danger of incidents that are related to air traffic control service varies with how close the planes come to each other and the time margin for evasion instructions and action. Thus, the 'serious' and 'considerable' degrees of danger apply to all of the following examples:

1. Failure to notice that two planes had come close until after they had passed each other.
2. The problem was first noticed by the controller when a pilot reported it.
3. When the problem was noticed, the two planes were already extremely close and there was no time to give evasion instructions. The reason that the incident did not develop into an accident can only be said to be evasive action by the aircraft crew or 'good fortune'.
4. The result is that two planes are on the same runway or on cross-runways at the same time.
5. Some last-minute instructions for evasion could be given, but the planes came extremely close to each other.

If we look at the proportion of incidents for each type of service for which the degree of danger was 'serious', that proportion was highest for air route traffic control, and most often both of the involved planes were flying on IFR. Looking at those results in detail reveals that the number of incidents that involved two commercial airliners and the number that involved a commercial airliner and a military aircraft were about equal.

As shown by the expression 'the critical 11 minutes' (the three minutes after take-off and the eight minutes before landing), it has generally been recognized in the field of aviation that danger has a high rate of occurrence in the vicinity of airports. In Japan, there is concern for near misses involving commercial airliners and military aircraft near airports that are shared by civilian and military aircraft, such as Nagoya Airport, and this is seen as an occasional problem.

Also, it has been considered highly dangerous that military aircraft fly under visual flight rules. As shown in the discussion above, however, the results of analysing the 545 incidents make it clear that there is quite a difference between what is generally recognized and the way things actually are.

Close encounters of aircraft can be broadly classified into those in which the aircraft are flying in the same direction and those in which they are flying in opposite directions (flying head-on). When planes that are flying in opposite directions approach each other, the distance between them shortens faster than when they are flying in the same direction, and, if the air traffic controller is delayed in noticing the situation, there will be no time to issue evasion instructions, so the degree of danger can become high in an instant. Accordingly, a high proportion of incidents that involve traffic moving in opposite directions have a 'serious' degree of danger.

Furthermore, as described earlier, there was a strong recognition that two planes are generally involved in near miss or confliction. It is clear, however, that consecutive near misses and conflictions by aircraft that result from instructions to a plane to avoid collision with another plane involve a high degree of danger. This is also true in the case of maritime accidents, as shown by the accident in which the Japan Self Defence Forces submarine Nadashio collided with a fishing boat when the submarine was manoeuvring to avoid a yacht near the Miura Peninsula on 23 July 1988.[2]

3.18 Urgently needed improvements

When we further considered the results of analysing individual incidents three-dimensionally with respect to the three classes (en route/terminal/ tower) of air traffic control services using quantification method III, we were able to identify a number of typical incidents in the same way as was done for the pilot-related incidents. While I omit the details here, we obtained the following results by examining the factors that have a high degree of danger in the dimension where the factor correlations are strongest for each of the types of air traffic control service.

Special flights have a high correlation with the 'serious' degree of danger. Also, although there is some difference among the types of control service (the difference between 'serious' and 'considerable'), the factors that are associated with a high degree of danger are controller in training and, closely related to this, the workplace environment, controller unfamiliarity, and insufficient controller training and practice (particularly at times of accidents and unexpected radar malfunction), as well as pilot requests, slow action by pilots, miscalculation of aircraft climb/descent rate and turning circle, insufficient co-ordination, shift change, and so on.

It has also been verified that, in every case, the information processing phase in which the incident was determined (i.e. the incident critical phase) is the decision phase. In air traffic control service, decisions are nearly always directly connected to instructions, and, except for cases that involve misspeaking or unclear pronunciation, the information processing error mode flow most often progresses through decision and instruction, the same categories as described in Chapter 1.

That the decision phase is the incident critical phase reveals a characteristic of air traffic control service and can be said to be an appropriate result.

If we look at the error modes that are associated with a high degree of danger, what correlates comparatively and relatively with 'serious' is vacillation in the decision phase and delay in the action/instruction phase. The reason for this is that aircraft that are not on the ground cannot stop in midair, and so, if for any reason the air traffic controller hesitates in making a decision and is slow to issue instructions, the aircraft continue to fly at high speed and their positions with respect to each other continue to change, steadily narrowing the distance between them and increasing the possibility of a near miss.

There are large disparities in the relations between error mode and degree of danger among tower control, terminal radar control, and en route control in delay in the information receiving and decision phases, excessive focus in the information receiving phase, and 'fixation' in the decision phase and the action (operation) or instruction phase.

Tower control and terminal radar control are positioned close to the 'serious' degree of danger, indicating a close relationship, whereas air route control is positioned close to the 'virtually none' degree of danger.

The reason for this difference is considered to be that the Radar Data Processing System (RDP) used in air route control is equipped with a warning device. The operation of this warning device serves to draw the controller's attention to the dangerous situation when aircraft approach each other while the controller's attention is focused on other aircraft. This indicates that we can reduce this kind of danger by developing warning devices for the Automated Radar Terminal System (ARTS) and the Terminal Radar Alphanumeric Display system (TRAD) that are used in terminal radar control and installing those devices in all terminals.

Chapter 4

Aircraft maintenance personnel and latent danger in aircraft

4.1 Significance of maintenance work and associated stages

According to the Convention on International Civil Aviation (the 'Chicago Convention'), 'Every aircraft engaged in international navigation shall be provided with a certificate of airworthiness (certifying that the aircraft satisfies certain technical standards for strength, structure, and capability) issued or rendered valid by the state in which it is registered' (Chicago Convention, Article 31). In Japan, this certificate is issued by the Minister of Land, Infrastructure and Transport, and no aircraft may be in service unless it has a valid airworthiness certificate and no aircraft shall be in service beyond the scope of its purpose or operating limitations as designated in the airworthiness certification (Article 11 of the Civil Aeronautics Law of Japan).

Even aircraft that have been certified as airworthy, however, decrease in airworthiness as time passes and through use. Therefore, maintenance, repair, and modification work is done to maintain airworthiness or to improve the quality of the aircraft equipment. When such work is done, it is checked for conformity with technical standards for ensuring safety that satisfies the airworthiness certificate (Article 19, Para. 1 of the Civil Aeronautics Law of Japan). If the work does not pass inspection, the plane cannot be flown (Article 16, Para. 1). Accordingly, inspection, testing, and checking are important tasks in maintenance work in addition to maintenance, repair, and modification.

For that reason, the incident reports that involve aircraft maintenance fall into two types: those that describe problems that occurred as the result of the reporter's own errors during the actual maintenance work and those that describe problems that relate to the mistakes of others during inspection, testing, or checking work or the discovery of problems due to deterioration over time. In the following discussion, I refer to the former type as 'occurrence' and the latter as 'discovery'. The difference between occurrence and discovery, as described later, has an important significance (for reference, this distinction is noted in each of the cases that are presented).

The maintenance work that is performed is determined by time period and other conditions, and varies with the airline company and the type of aircraft. These are called the maintenance stages, and there are generally six of them: T, A, B, C, shop, and special (H and big modification, etc.).

Here, as an example, we will have a look at the Boeing 747 maintenance stages of airline A:

1. *T maintenance*. This is known as the preflight inspection. Within the short time between the plane landing and the next flight, a check is made for any problems that would impede the following flight, to maintain safe operation. When operation begins, T maintenance must, by law, be performed, and must be completed just before the plane departs. Accordingly, it is performed for each flight of the aircraft.

2. *A maintenance*. This is the condition check (an inspection to determine if the aircraft can tolerate continuous use), periodic maintenance that most directly affects operational safety. It is performed at least once every 300 h.

3. *B maintenance*. This maintenance stage is positioned between A maintenance and C maintenance. It involves inspections that are performed at specific intervals and maintenance (mainly repairs). It is performed at least once every 1200 h.

4. *C maintenance*. This is a higher stage of maintenance than A maintenance and B maintenance. It involves preventive maintenance on functional systems that have a high possibility of affecting aircraft operation and preventive maintenance on relatively easy-to-access airframe structures (checking and inspection). It is performed at least once every 3000–4000 h.

5. *Shop (factory) maintenance*. This maintenance is performed by taking parts or equipment off the fuselage or engine and sending them to the factory for work.

6. *Special maintenance*. This is special work that is done according to a modification order, 'H maintenance', and so on. H maintenance is performed as needed, and not with a particular frequency or at

regular intervals.   It is performed according to the cumulative necessity or urgency of repairs or modifications that are difficult to do on a daily basis.   This maintenance involves work that takes a relatively long time to complete.

The place at which maintenance work is done is different for the different stages.   T, A, and B maintenance is generally called line (operation) maintenance.   It is usually performed on the ramp or spot. The ramp is the place where the plane is ordinarily parked or tied down; the spot is where passengers embark and disembark.   C and H maintenance, on the other hand, is normally done in the maintenance area (hangar).   Shop maintenance is done at the factory.

It might be said in passing that maintenance done on the spot or ramp is done out of doors where there are no fixed facilities for maintenance work, but, in the maintenance area (hangar), ceiling lights and movable work platforms for various types of aircraft are available.   The shop (factory) provides a better work environment for maintenance than the other maintenance locations, as well as testing equipment and suitable tools.

4.2 Focus on lack of sleep

Concerning the background factors of unsafe events that are related to aircraft maintenance, various factors were discovered in the same way as for the pilot-related factors described in Chapter 1.   Those factors include personal factors (factors related to knowledge and experience, physical factors, and psychological factors), interpersonal relationship factors, general factors, special background factors, and so on.   For personal factors and interpersonal relationship factors, in particular, many of the factors are the same.   Here, I will explain only the factors that deserve particular attention among the factors that are related to maintenance work.

Of the personal factors, one demanding particular attention is lack of sleep.

Human physiological functions change according to what is practically a 24 h rhythm, which is known to affect human activity in various ways.   Because line maintenance is performed on a 24 h basis, the work goes against the body's natural rhythms, resulting in lack of sleep, which competes with fatigue and darkness to become a background factor for errors.   The time of occurrence is most often at twilight.   The following is an example of that.

An inspector was inspecting the passenger cabin floor beams

(reinforcement materials that support the floor of the fuselage) during H maintenance on the night shift in the dawn hours of about four o'clock. He was suffering from lack of sleep and it was difficult to maintain the strain owing to inattentiveness after the stress of a night's work. Because the lighting was dark, he failed to notice cracks that could have been seen during daylight and finished the inspection in an easy-going manner. The erroneous ending of the inspection was discovered by the inspector of the next work shift, on the day shift. The degree of danger was 'fairly little' in this example (occurrence).

The reporter of this incident wrote that 'When working the night shift with no nap, the sleepiest time is at dawn, and detailed inspections are difficult, so I think that we should avoid doing such inspections at that time'.

The difficulty of visual inspections is pointed out fully in the report of the US National Transportation Safety Board, which investigated the accident in which a part of the passenger cabin ceiling (upper part of the fuselage) of Aloha Airlines flight 243 Boeing 737 bound for Honolulu was suddenly blown off in midair at 24 000 ft, resulting in cabin decompression, on 28 April 1988.[1] The accident is inferred to have occurred because an outer plate ripped off when a fatigued lap joint (joint between outer plates) rivet broke. This cracking was not discovered, even though it could have been detected by the inspection by eddy-current testing that was performed the previous year.

The inspector must attach a safety rope so as not to fall off the fuselage, hold a light in one hand while climbing up on a narrow scaffold, and move along the top of the fuselage from one stepping place to the next. Even assuming for the moment that the temperature and lighting conditions are suitable, examining hundreds or thousands of rivets one at a time for minute cracks is extremely monotonous work and, at the same time, extremely dangerous work. When a certain number of rivets have been inspected without finding cracks, the inspector might begin to think that 'There probably aren't any cracked rivets after this, either', and that would only be natural.

If the company would give sufficient consideration to the length of working hours, the single-person working environment, how tired the inspector is, lack of sleep, irregular work, rest schedule, work time deviation from the 24 hour biological rhythms, and so on, then it would probably be possible to perform more effective inspections.

The US National Transportation Safety Board report, in pointing this out, identifies the unreasonable maintenance programme of Aloha Airlines as the cause of this accident.

4.3 Poor team co-ordination

When a number of maintenance workers are performing the same work at the same time, the workers not sharing the same understanding of a situation and not working in synchronization become background factors to incidents.

An example of this is an incident (occurrence) of 'considerable' degree of danger in which an Auxiliary Power Unit (APU) was being hoisted with two power cranes in front and one manual crane behind because sufficient equipment was not available. A metal segment of the APU broke in the final stage when it was being hoisted in an unbalanced state because of lack of effective teamwork between the two workers who were operating the power cranes and the manual crane.

Because the footing was not good, the maintenance person who was watching the operation could not stand in the most suitable place, and, although the person who reported the incident knew that the unit being hoisted was not level, he could not see the upper part of the unit. As a result, it was difficult to judge the possibility of it hitting the fuselage of the aircraft. Another background factor in this case was darkness.

There is also an example of a lack of effective teamwork resulting from different understandings of the objective when people are working at the same time on a single objective.

Two maintenance team members were performing a manual operating test on a spoiler. One of them was about to raise the crane to remove a brass rod that had been inserted temporarily, but the other team member thought that the crane was being raised because it was close to the hinge.

The first team member, thinking that the brass rod would be easier to remove if the crane were raised further, raised the crane again, exceeding the load capacity of the rod and breaking it. The degree of danger was 'fairly little' in this incident (occurrence).

4.4 Incidents related to work platforms

To reduce air resistance during flight, aircraft are streamlined, and, in contrast to ordinary buildings or structures, there are almost no places to step on the body of the airplane itself. Commercial airliners are large and quite high, so, except for a small part of the maintenance work (cockpit instruments, passenger cabin seats, etc.), work platforms are needed.

Work platforms are important when performing aircraft maintenance, inasmuch as the first thing that must be done is to prepare a work platform that is suitable for the work. Hangars are equipped with movable work platforms, but, for T maintenance and other types of maintenance that are performed outside the hangar, portable work platforms or stepladders are used. There are also cases in which the climbing or descending of work platforms, their movement, their setting in place, or insufficiency has been related to the occurrence of incidents, or in which aircraft have been damaged when a work platform has struck the plane while being moved into place. The degree of danger in such cases is also high.

Even in the case of maintenance performed inside the aircraft without the use of work platforms, there are surprisingly many cases of insecure footing because the floor surface is not necessarily level or is slippery. For efficient maintenance work, firm footing is a necessity. When footing is insecure, excessive effort is required to maintain one's balance, leading to background factors such as excessive or reckless action, easy-going decisions, and failure to notice incorrect actions. Also, loss of balance or stability can result in a violent fall or drop and the overturning of a toolbox or other object by the worker's foot, resulting in damage to the aircraft.

The following are examples in which the background factors include insecure footing or insufficient equipment during maintenance work.

In one incident (occurrence), there was no suitable work platform when a spoiler support was being replaced, and, although the worker was wearing a safety belt, he slipped on the inclined surface of the wing. His body weight was shifted onto the hand with which he was supporting the spoiler and he lost balance, breaking the temporary support for the spoiler and resulting in damage to the trailing edge of the spoiler. The degree of danger was 'considerable' in this case.

The following is another example (discovery).

The work was being done in a dark and tight freight compartment that had a slippery fibreglass floor. Because it was difficult to see and the work was done by groping with the hands, it was discovered during the inspection that the safety wire had been applied in the wrong direction because of the worker's body position. The degree of danger in this case was 'fairly little'.

Incidents (occurrences) such as the following have also been reported.

A stepladder that was used to replenish the lubricating oil of the engines of a Boeing 727 was left under the No. 3 engine, as the worker was going to clean up afterwards. While he was gone, baggage and freight were unloaded from the plane, which caused the tail end of the plane to sink lower. Because of that, the stepladder was caught between the engine and the ground, causing damage to the engine cowling (the outer covering of the engine). This is an example in which the degree of danger was 'virtually none' and the background factors included insufficient experience, work interruptions, and work platform.

The centre of gravity of a plane changes when baggage and freight are loaded and unloaded, but how much it changes and whether the fuselage drops or rises can only be known through experience.

The three engines of the Boeing 727 are all mounted on the back part of the fuselage, and, for front–back balance over the main landing gear, nearly all of the passengers and freight are loaded forward of the main gear. Because of that, the front end of the plane sinks lower as freight and passengers are loaded, and rises when they are unloaded. The maintenance worker appropriately understood that it was necessary to take away the stepladder and that luggage was being unloaded, but he did not believe that the tail of the plane would come in contact with the stepladder, and sat idle with an easy-going decision.

4.5 Work difficult to see under dark conditions

In dark places, even people who have good vision do not see well, and there is a tendency towards near-sightedness (referred to as 'night-time myopia').[2] For that reason, it is easy to fall into errors of inappropriate action or mistakes on account of failure to see, or seeing incorrectly, and thus making easy-going decisions or having wrong convictions. When these are compounded by physiological factors such as fatigue or lack of sleep, and even further by a tight work environment, the possibility of making mistakes is even higher.

To avoid this kind of danger, adequate lighting is required above all else, but, even in the hangar, it is difficult to provide suitable lighting from all possible directions because of the installation of work platforms, etc., and darkness contributes greatly to errors in aircraft maintenance.

The times of day in which incidents have darkness as a background factor are night, dawn, and twilight, but such incidents also occur during the daylight hours. They also occur even in the hangar, where lighting is not limited to lamps and spotlights.

Many incidents that have darkness as a factor are related to engines and landing gear. Engine maintenance involves a large amount of work, but, even with the engine cover completely opened, it is difficult to light up the engine from all sides, and the fact that much of the work is finely detailed has an effect. In maintenance of the landing gear, it is difficult to obtain suitable lighting because of the effect of the fuselage and the work platform. Another problem is that parts stained with oil are difficult to see. Also, darkness inevitably conflicts with a tight working area.

The following incident (occurrence) occurred at dawn on the night shift, with the inspection being mainly line maintenance.

One of the wheel tie bolts (bolts that fasten the foil part of the wheel) was not tightened properly and was loose, but, because it was hard to see, it was considered to be all right and passed inspection. After being put into service, it was discovered that a bolt was missing during the runway check at the airport. The degree of danger was 'considerable', and the reporter of the incident received 'severe admonition'.

As described by the incident reporter:

Checking is the main purpose in line maintenance, but, because it is often done at night when the darkness makes it difficult to see well, many mistakes in checking are made. Performing this work when one's visual acuity is low (well known from vision testing at dawn on the night shift) in a limited timeframe wears one's nerves thin.

The ability of night-shift workers to see in the dawn hours is considered to be reduced by fatigue and lack of sleep.

The following is a typical incident (discovery) in which the darkness and tight workspace that are characteristic of aircraft structure play a role:

The problem was that the readings on the No. 1 (pilot-side) compass and the No. 2 (copilot-side) compass (which indicate the aircraft's heading) differed by more than 20°. Prior to this, there had been frequent compass-related problems, and various instruments and parts had been replaced. Recently, there had been quite a few problems, so a detailed investigation was done.

When the parts and wires were checked scrupulously, one by one, a defect was finally found. According to the report, the cause of the problem was a connector of the flux valve (the electromagnetic flux detector) of the No. 1 compass. When the three-wire shield (a wire that converts the compass heading to an electrical signal and carries that signal) was connected to the pin, one wire of the shield pierced the insulation of one of the three wire cores. This probably resulted when the shield was turned back improperly at the time this wire was replaced, and the fact that the wire insulation had been pierced as mentioned above was not noticed.

The turned-back part of the shield was soldered over, so, no matter how much attention was given to it, I don't think it would have been noticed. What is more, in addition to the darkness around the flux valve, the wire had little excess length to work with, making the work extremely difficult to perform.

From the results of operating tests, also, it is difficult to judge that the flux valves, which are installed in the wing tips so that the pilot-seat compass can display the magnetic heading, are not operating normally, and they have mistakenly been passed as operating normally.   The reason for recovery was 'pointing out by the crew' and the degree of danger was 'considerable', because the compass is basic to navigation.

4.6 Mistakes due to tight spaces

Commercial airliners are steadily increasing in size, and the general image of largeness is strong. In civil aviation, however, the requirement is to transport passengers and freight speedily and in large volume to the destination, and to do so at as low a cost as possible. Accordingly, one of the important design policies for commercial aircraft is to make the payload space as large as possible. The various aircraft equipment is made lighter and more compact, and, to save space, it is inevitably concentrated in small areas, thus making maintenance difficult.

Tight workspace and darkness are closely related, and, when tight workspace is a background factor, darkness is also often involved. That is not always the case, however; other factors may also be involved. Tightness of workspace easily leads to inability to see clearly, inability to reach sufficiently, failure to notice, recklessness or forceful actions, and insufficient or incomplete work.

Next I will present an incident (occurrence) in which the No. 1 (pilot-side) radio altimeter failed to work, because of faulty maintenance work done at an earlier time. The maintenance involved crimping together the three wires of the radio altimeter in the extremely cramped space behind the pilot's instrument panel.   However, the crimping was insufficient and incomplete, and, as a result, the wire loosened and the problem appeared some time later. The problem was identified by 'pointing out by the crew' and the degree of danger was 'fairly little'.

I removed the front pilot's instrument panel and then reinstalled it. At that time, I reconnected a number of wires that I had cut for convenience in performing the work, including the radio altimeter wires.

Because the wires were cut at the junctions, I reconnected them by the same method. As you know, however, the space behind the instrument panel is very tight, and, although I did the crimping together of the three wires of the radio altimeter in a manner different from the usual, those, too, were connected by the same method as they were originally.

No problem appeared in the operation testing after the instrument panel had been reinstalled, so the maintenance work was ended. Recently, however, when over a year and a half had passed since the maintenance, the No. 1 radio altimeter stopped working. Although various components of the system were replaced, including the transceiver, the indicator, and the antenna, it made no difference. Meanwhile, the symptoms stopped appearing.

With nothing else to do, the system was checked while disconnecting the wires that connect to the instrument one by one. In that way it was confirmed that the same symptom appeared when the No. 6 pin was disconnected, and tracing that wire led to the connection point behind the instrument panel. While from the look of it there was no apparent problem, it was possible to reproduce the symptom if that wire was shaken and the wire came off with a tug. After the wire was reconnected, the problem did not recur.

4.7 Dangerous work within fuel tanks

Most of the commercial airliners have fuel tanks mounted in the wings. For strength, the wings are constructed with many internal ribs and bulkheads, and many fuel tanks are arranged between those structures. Each tank has a fuel probe for measuring the amount of fuel that remains in the tank, and so the wiring for all those fuel probes runs through the wings.

When there is a problem with any part of this system, the maintenance worker must access the area through an access door in the surface of the wing. To maintain the strength of the wing and to reduce air resistance, the wing surface must be as smooth as possible, and so these access doors are very small. On a Boeing 747, for example, the left and right wings each have 25 access doors, but the doors are elliptical in shape and are 10 in wide and 16 in long, the minimum size that allows a person to enter and exit.

Also, because the wing interior is filled with toxic fuel vapours, the worker must wear dust-proof goggles and a mask to which an air line is connected so that outside air can be breathed. Moreover, the space inside the wing is very tight, and working there is difficult in the extreme. In the summer, the temperature inside the wing can rise to over 40° C (104° F), and is increased further by the heat from the hand-carried work light. Sweat fogs the worker's goggles and it is difficult to see. This is an extremely poor working environment.

To inspect the inside of the fuel tanks, removable ribs (removable structures that support the outer plates of the wings) are taken out when entering the tank. When exiting, those removable ribs are returned to their original positions. There is an incident (occurrence) in which a wire was jammed under the head of a bolt when a removable rib was put back in place. The jammed wire was a low Z wire, which is required for displaying the amount of fuel remaining on gauges in the cockpit.

It is necessary that a certain amount of fuel always remain in the front part of the wings during flight, so as not to exceed the critical strength where the wings are attached to the fuselage. A Boeing 747 has about 30 fuel tank units in each of the left and right wings; the flight engineer must pay attention to the amount of fuel remaining in each of those tanks and maintain a balance among them during flight.

If the readings on the cockpit gauges for remaining fuel are inaccurate because of a defect in the low Z wire or other such reason, the fuel balance will be lost, creating the possibility of the strength limit of the parts where the wings attach being exceeded. For that reason, although the electric current flow was very small compared with the current of a motor, the degree of danger for this case was 'considerable'. The recovery was due to 'operation testing'.

The reporter of this incident wrote: 'Because the work was done in a tight space while lying on my stomach and the bolt was inserted by groping with the hands, I did not look upwards afterwards'. The reporter of another incident that involved work done inside a fuel tank also wrote: 'All I could think about was getting out of the tank quickly, so I didn't look back with sufficient care'. From this kind of information, too, we can understand the difficulty of working inside these fuel tanks while wearing a mask.

4.8 Danger due to inability to make direct observations

There are many places in an aircraft that cannot be seen directly after assembly but nevertheless also require maintenance. The following case (occurrence), which involves a problem that occurred because the teeth of the gears in the flap driveshaft coupling (a joint in the shaft that moves the flaps) cannot be seen from the outside, is an example.

Because the safety wire was securely fastened, the maintenance member believed that everything was all right and that there was nothing to prevent departure, but, during the take-off climb, the flaps failed to operate and the plane had to turn back after take-off. The degree of danger in this incident was 'considerable'. In maintenance work that was done the previous night, the coupling had been installed with the driveshaft not meshed enough, but this was not known because the flaps operated properly on the ground. The coupling was disengaged because the wings were distorted by their weight in the air, but this had not been discovered because the coupling had operated without problems on the ground. Finally, it was understood that this maintenance work had to be redone.

4.9 Shortage and substitution of parts and materials as a cause of accidents

An aircraft is constructed of innumerable parts and materials. These parts include some that have predetermined limits on the periods of time over which they can be used, and that must be replaced before those limits are exceeded; others that are replaced or repaired as problems occur owing to deterioration over time; and still others that must be repaired because of damage caused by mistakes made in maintenance work or other work. The objective of aircraft maintenance is to maintain airworthiness, and the maintenance of airworthiness through the replacement or repair of faulty or damaged parts and equipment was described earlier in this book.

Most of the parts and materials are managed in the parts warehouse (store). Sometimes, however, a needed part is not available or there is not enough time to obtain the part from the store, and so a similar part or material that is on hand is used in place of the needed part. Parts are thus sometimes used with the knowledge that they are unauthorized because they seemed to work all right when tried and because of the customary knowledge that such substitutions are frequently made. In many cases, such substitutions seriously affect the time constraints on use of the part.

Although the part that is substituted for another part is not itself flawed or defective, problems may arise because of incompatibility with the system in which it is installed when used with a system other than the one for which it is intended (part substitution). Also, when materials that are prepared for a particular purpose are used for other purposes (material substitution), problems can be caused by differences in properties from the original material, improper amounts, and so on.

As to whether or not parts were available when the maintenance work was done, when the incident is reported by someone other than the person who did the work (i.e. the discoverer of the problem), it is often the case that the reporting person knows only that the part has suffered deterioration over time. That is to say, there is a possibility that, for discovery incidents, the incidents that have part deterioration over time as a background factor may include incidents for which part or material unavailability is a background factor.

In one such incident (discovery), a conduit that passes through a pressure bulkhead was reinstalled after having been removed for convenience in maintenance. Rather than preparing a new seal for the space between the conduit and bulkhead, one that had been shared with neighbouring workers was used (substitution). However, the sealing effect was insufficient and an air leak occurred during the leak test. The degree of danger in this case was 'virtually none'.

There is an example, albeit somewhat old, of an accident caused by the use of wire instead of the authorized part. On 4 October 1960, an Eastern Airlines Lockheed Electra 188A four-engine propeller aircraft crashed. The US National Transportation Safety Board, which investigated the accident, attributed the crash to the fact that the aircraft encountered a flock of starlings just after taking off from Boston Airport, which resulted in the No. 1 engine stopping and a temporary loss of power output in the No. 2 and No. 4 engines. The Board concluded that 'The considered cause of the accident was loss of engine power due to intake of wild birds and an inability to recover, an unprecedented and fatal sequence of events that led to loss of airspeed and control during the take-off climb'.

However, in the lawsuit for compensation by the families of the victims, the plaintiff's counsel continued to press for the true cause of the accident for 6 years. In the autumn of 1966, by a stroke of luck, they were successful in getting the highly suggestive opinion of a competent pilot who had abundant flight experience in the same type of aircraft.

That pilot first pointed out that one of the engines of the plane that crashed was working properly and that two others also immediately regained power. This type of aircraft is supposed to be able to maintain altitude with at least two engines functioning normally, so it is hard to see why the pilot of that plane would be concerned about speed of the aircraft because one engine had stopped. That is to say, the problem is why the pilot brought the nose of the aircraft up sharply.

Therefore, the counsel for the plaintiffs, according to that pilot's advice, requested to be provided with the maintenance records of the crashed plane and discovered from those records that there had been a problem with the copilot's seat. About 6 weeks prior to the accident, an incident had occurred in which the copilot's seat had suddenly slid backwards during a take-off, and the same thing had happened again during the landing of that same flight.

The maintenance member who had dealt with that problem could not replace the problem part because no authorized spare part was available, and so he used a wire of different length to the authorized rod instead. Moreover, it was confirmed that, even after this repair, the seat could not be fixed in the specified position and, even in the inspection that was done 2 weeks before the accident, the wire had not been replaced by the authorized part.

In the trial, the court accepted the conjecture that the copilot's seat had suddenly slid backwards during the take-off climb, causing the pilot to pull back sharply on the control column, with the result that the nose of the aircraft had gone up, reducing the aircraft speed. The conclusion of the National Transportation Safety Board was revised in face of the facts.[3]

This accident is a reminder of the importance of early replacement with authorized parts when non-authorized parts have been used as a temporary measure because authorized parts are unavailable. In any case, it is worthy of attention that the fact that Eastern Airlines kept the maintenance records and made them available was the basis for clarifying the cause of the accident.

4.10 Mistakes due to poorly labelled/indicated parts

Of the innumerable parts of aircraft, there are some that have exactly the same or similar shape but have different capabilities and cannot be distinguished from outward appearance alone, and others that have the same capabilities but do not look the same and are for different systems. For that reason, virtually all parts, including wires, are labelled with a part number or wire number. Thus, checking these identification numbers before performing maintenance work is fundamental to the job of maintenance.

However, these part number and wire number labels may become difficult to read because they wear off or become obliterated by dirt and grime, or because the part itself is small and the lettering of the label is small and hard to read (unclear label). In rare cases, there may be no label on the part to begin with (absent label).

There is an example incident (discovery) involving maintenance work done during the short time between landing and take-off, for which the degree of danger was 'fairly little'.

In a parts bin, there happened to be a part that had exactly the same external specifications as the part with which the bin was labelled, but that had a different part number. Although the maintenance worker fully understood the necessity of checking the part number, the part was small (a small light bulb) and so the part number was hard to read, and, as a result, he mistakenly installed the wrong part.

Because the bulb that was mistakenly used had a low rated voltage, the current flow was high and it burned out. The reason for recovery was 'pointing out by the crew'.

Prevention of the mixing up of parts that have exactly the same appearance but different capabilities and that furthermore have small, difficult-to-read labels probably requires an improvement such as packaging the bulbs individually in plastic mounted on cardboard labelled with the part number in large letters at the time of manufacture. Fundamentally, however, parts should be designed such that only the prescribed parts can be installed in a given location.

4.11 Pitfalls of shift changes and work interruptions

Aircraft maintenance work is performed over two or three work shifts. Because of that, it is usual for shift changes to occur while work is in progress. Also, it is sometimes necessary for the person responsible for some maintenance task to turn the work over to another worker before it is finished because another task requires his attention.

When a shift change occurs while a task is in progress, the work remaining to be done is described on a work transfer form so that the work can be continued on the next shift. Sometimes, however, part of the information is passed by word of mouth, and it may happen that the meaning is not fully conveyed or some task items are left out owing to a lapse of memory.

It is also necessary, at times, to interrupt work temporarily for various reasons, such as meals and work breaks or to get parts from the parts store. When such interruptions occur, the worker's mind is temporarily taken off the task, and so, when the worker returns to work, it is easy to have forgotten where he left off with the task, to have mistaken convictions concerning changes in the work situation during the interruption on account of preconceptions, or to make errors in judgement owing to failure to notice changes that have occurred since the work interruption.

Shift changes while a task is in progress and work interruptions have one point in common: discontinuity in the worker's awareness and understanding of the work circumstances. The following incident (occurrence) is one case in which work interruption alone is the background factor.

The reporter of the incident was tightening the pipe of a P_{S4} while engaged in preparing an engine for operation after cleaning, when he was asked to help with another task. (The P_{S4} is a sensor that detects internal engine pressure for a controller; if this part does not function properly, the engine does not develop thrust.) Afterwards, he went to retrieve a part and forgot about tightening the pipe.

Because the pipe was somewhat tight, it was difficult for the inspector to discover that the pipe had not been completely tightened by touching with his hand. There was also no problem apparent when the engine was test run to dry it out after the water had been drained out, so the maintenance work was finished and the aircraft put into service. However, during the take-off run of the second flight of the aircraft, the engine lost power and the take-off was rejected. The reason for recovery was 'pointing out by the crew' and the degree of danger was 'considerable'.

In the following case, work interruption and insufficient communication, and complicated or parallel work, painting, are the background factors to the incident (discovery).

The maintenance worker, who was doing departure preparation maintenance on the first departing flight at dawn, discovered holes in the underside of the right wing, forward of the outboard aileron. A closer look from up on the work platform revealed them to be the holes for the rivets that fasten the hinge of an inspection port panel to the wing. Seven missing rivets were replaced and the plane was allowed to depart.

In the C maintenance that was done up to the previous day, two repair instructions were issued, one for corrosion on the panel and the other for replacement of the hinge. First, the hinge replacement work was done. The old hinge was removed, but the corrosion of the panel surface had proceeded to the area beneath the hinge. Because the hinge rivets could not be applied unless that corrosion was removed, the new hinge was attached temporarily with pins and the work interrupted.

The maintenance worker that did the corrosion removal removed the corrosion from the parts of the panel and completed the task according to instructions, but, because the hinge was a separate job, he left it as it was. After that, the panel was temporarily closed for the purpose of painting of the area around it. After the painting, the panel was left closed.

The next day, the job of installing the hinge was continued by another person. When he approached the work site, that worker saw that the panel was closed and got the mistaken impression that the job had been finished and the previous worker had forgotten to sign off on it. He then signed the work completion form himself and handed it in. Thus, according to the records, the work had been completed.

The degree of danger in this case was 'considerable'.

Incidents (occurrences) such as the following have also been reported.

Because I had taken a break and was feeling rather relaxed, I forgot to connect the duct (that carries pneumatic air to the motor of the thrust reverser), which had been detached for boroscope inspection of the interior of the engine, and closed the engine cover. Without discovering the mistake, the aircraft was put into service. Right after take-off, the engine fire alarm turned on and the plane made an emergency landing'.

The degree of danger in this incident was 'serious'.

The reporter of the incident said, 'Now, it's just a matter of common sense, but that type of aircraft had just been introduced', and unfamiliarity is a background factor. Forgetting to connect this duct does not simply prevent the thrust reverser from working; if there is also a fuel leak, the possibility of a fire occurring is high. Interruption of work by a rest break, too, is a major background factor that leads to forgetting. A means of reminding workers so that they do not forget to complete the previous task, such as attaching a tag, is probably required.

In another incident (occurrence), the fuselage that required maintenance was mistaken because of a preconception due to a work interruption. The maintenance worker began his break after confirming that the aircraft that was to be repaired was parked at the spot. However, when his break was over, a different plane was parked at that spot, but the worker mistook that plane for the one on which the repair work was to be done, and so replaced the part on that plane. The plane that the worker had confirmed to be at that spot had been moved to a different spot during his break, and then another aircraft of the same type was later parked there.

There is also an incident (occurrence) in which a maintenance worker confirmed the circumstances of the aircraft before beginning the work to determine the parts needed and then went to the parts store to get the parts. Although it is rare for two aircraft of the same type to be parked next to each other, in this case there were. Getting the replacement parts took more time than expected, and, because the worker had to hurry because of time constraints, he mistook the two similar aircraft and replaced the parts in the wrong one.

4.12 Danger caused by insufficient communication between personnel

When a shift change occurs while work is in progress, when a number of persons are working together on the same task and one of them interrupts his work, or when work is being done in parallel on different sections of the same system at the same time, the workers involved must share a consistent understanding of the circumstances of the work. The lack of such a shared understanding is a factor in the occurrence of unsafe events.

Because this is so important in the field of maintenance, a basic method for accomplishing it should be established. However, if, in the communication between maintenance workers, information is left out because the person who is turning over the work believes it is obvious to the person who will continue the work, or if the description is general and lacking in specifics, the meaning may not be completely understood or a meaning other than what was intended may be understood.

This sort of miscommunication is frequently experienced by all of us in our daily lives, particularly when a shift change occurs while work is in progress. For example, a worker may choose to continue working up to a point where it is convenient to stop, leaving insufficient time for the work transfer description (which should be in written form), and so the information may not be conveyed fully.

In cases such as when the maintenance involves work being done in parallel on different sections of the same system at the same time, too, an extremely dangerous situation can develop owing to inadequate communication.

An aircraft has various power supply systems (electrical, hydraulic, and pneumatic), and, for example, if a hydraulic system pipe is detached for one maintenance task while another maintenance task requiring that a hydraulic pump be operated is being done, oil will be expelled from the detached pipe unexpectedly. As another example, if work is being done on the flaps, and, when unaware of that, other maintenance personnel turn on the power in the cockpit for other maintenance work, the person who is working on the flaps may be crushed to death between the flap and the wing. (There have been fatal accidents in which this actually happened.)

The valves of hydraulic systems and pneumatic systems are nearly all controlled electrically. Accordingly, most systems will not function if the electric power is turned off, even though the hydraulic pressure or pneumatic pressure is up. However, if the power is turned on by mistake when the pressure is up in a hydraulic or pneumatic system, that system may suddenly operate. This is an extremely dangerous situation.

Careless actions with respect to power systems can result in a high danger of accidents that cause bodily harm to people as well as damage to the aircraft. Because commercial airliners are large and it is difficult for all members of the maintenance team to know the work circumstances of the entire plane, in cases where harm will come from turning power on during maintenance work, 'Do not operate!' tags are placed on the relevant switches or levers before the work is begun.

One of three maintenance workers who were working together on the same task had nearly finished replacing a tyre and was tightening a retaining bolt when another job came up, and so he said simply to another member, 'Please look after it'. However, because of that vague and non-specific instruction, the intention of the person handing over the work was not fully conveyed and the maintenance person hearing it believed that the retaining bolt had been tightened and installed the cover over it. Twenty days later, it was discovered that the bolt had come out and broken through the cover.

The degree of danger of this incident (occurrence) was 'fairly little' and shift change during work and darkness were also background factors.

There are also examples such as the following.

In this incident (occurrence), a maintenance worker in the cockpit turned on the inverter switch, which turns on a device for changing the frequency of the power supply, without informing another worker who was in the E/E compartment because he was excessively focused on the fuelling and had forgotten that the other worker was there. (The E/E compartment is a room that contains the radio equipment of the aircraft and other electronics, in which the temperature must be controlled to a certain temperature or lower. It is usually behind and below the cockpit.) The result was that the other worker suffered an electrical shock because he was sweating as a consequence of working in the parking apron under the hot summer sun.

The degree of danger for this incident was 'considerable' and it occurred because of insufficient communication, although sweating was also a contributing factor. Sweating during maintenance work not only causes electrical shock, as in this example, but also, through sweat dropping on metal, causes corrosion. Sweating is one physical background factor that is peculiar to maintenance work.

The following sort of cases (discovery) related to the life of maintenance workers have also been reported.

A maintenance worker who was about to begin work on the stabilizers simply checked to see if the stabilizer hydraulic shut-off valve was in the closed position and did not place 'Do not operate!' tags on the hydraulic valve switch and electrical circuit breaker in the cockpit. Later, another maintenance worker switched on the hydraulic valve switch in the cockpit in accordance with the instructions for another maintenance task, not knowing what kind of work was being done on the stabilizers. Because the switch was turned on, the hydraulic valve opened automatically and the rudder moved to the right, hitting a work platform and damaging the rudder.

If a worker had been on the platform, the accident may have caused an injury, and the degree of danger was 'serious'. The background factors included insufficient communication, insecure footing, work platforms, insufficient supervision, and complicated or parallel work.

4.13 Incidents related to painting and polishing

One maintenance task is the periodic polishing of the outer surfaces of aircraft. This is done both to lower air resistance during flight and to prevent corrosion from the sulphides produced by corrosive gases in the air, such as sulphuric acid gas, and dust. Commercial airliners are also painted, both to protect the outer surfaces and to display the emblems of the airline companies, and must be repainted periodically.

Aircraft have static ports (holes) arranged with left–right symmetry on both sides of the fuselage for sensing the static air pressure (atmospheric pressure) during flight for the barometric altimeter, rate of climb indicator, airspeed indicator, and central air data computer (a computer that processes the data required for navigation, such as airspeed, altitude, and external temperature, and corrects deviations). There are also small openings for releasing excessive air pressure in order to adjust the internal cabin pressure, which are usually placed near the oxygen tanks.

When the outer plates of the fuselage are polished or painted, protective tape (masking tape) is applied to prevent these openings from being clogged by aluminium dust or paint from the polishing or painting and peeled off after the work is done. Most often, the taping, painting or polishing, and peeling off of the tape are done by different persons. During the painting or polishing, the masking tape, too, is covered with paint or aluminium dust, making it difficult to see where tape has been applied.

Problems such as instruments displaying incorrect values and permitted pressure ranges being exceeded during flight have occurred because workers forgot to peel off the masking tape before finishing painting or polishing maintenance and the mistake was not noticed during the inspections.

Also, paint prevents the flow of electricity, so, if masking tape is not applied to conductive parts before painting, the resistance of the part to electrical current will be increased by the paint, causing various problems.

During flight, aircraft take on an electric charge from the static electricity in the air. To release that electric charge into the air, pin-shaped static dischargers that are 4 in long and 0.4 in thick are installed on the trailing edge of the wing tips. If the part where the static dischargers are attached is painted, the electrical resistance will be higher and the discharger will not be conductive, resulting in insufficient electrical discharge.

If the aircraft becomes electrically charged, it greatly affects communication devices, and the shortwave radio and Automatic Direction Finder (ADF) in particular. There is also a danger of spark discharge when the plane lands. Another possible problem, albeit not so probable, is that parts that have a high resistance can melt when the plane is struck by lightning, reducing structural strength.

Because painting and polishing are, in nearly all cases, subcontracted work, the incidents involving this maintenance work have all been reported by the persons who discovered the problems.

In one incident (discovery), the static discharger was installed without removing the paint from the place where it is attached to the wing. As a result, the conductance test showed that the limit of $0.1\,\Omega$ was clearly far exceeded, with readings of about $60\,k\Omega$ at eight places on the left outboard aileron. The limit was also exceeded on the right outboard aileron in the same way.

From the maintenance record of the aircraft, this problem was inferred to have existed from the time of manufacture (a manufacturing error). It is necessary to remove the paint from the place where the static discharger is attached and to pressure-bond the discharger to the wing with a conductive adhesive. However, the worker at the aircraft manufacturer probably gave insufficient consideration to the fact that, if the static discharger was attached to a painted part, the static discharger would not discharge static electricity properly, thus causing problems such as hinges or other such parts being welded together by electric discharges when the plane is struck by lightning, so that the ailerons or other parts do not function. One wonders if the aircraft manufacturer fully understood the function of the static discharger. The degree of danger for this incident was 'fairly little'.

In another incident (discovery) for which the degree of danger was 'serious', the take-off was rejected because the airspeed indicator was operating in a strange manner owing to tape covering the static ports. The static ports have an important function in determining information that is necessary for aircraft operation, including altitude, speed, climbing rate, and pressurization. When the plane was repainted, the masking had been applied very smoothly, and, when the tape was covered with paint, it was nearly unrecognizable and passed through both the maintenance inspection and the preflight check (operation side).

Another incident (discovery) that involves painting is as follows.

The schedule for H maintenance was greatly delayed, and so the painting maintenance, which should have been completed earlier, was delayed. Because of the time constraints, the final inspection was done before the paint had completely dried. At the time of the inspection, the movement of the passenger cabin door lever was smooth, so the inspectors did not give sufficient consideration to the fact that it would not have moved after the paint had dried and so it passed the inspection.

The paint was thicker because of the repainting, so that the range of motion was reduced. It was discovered that the lever would not operate during the preflight preparation for the first flight after maintenance, and so the plane was not put into service. Therefore the degree of danger was 'virtually none'.

4.14 Danger caused by dust

Maintenance work often produces dust. Because aircraft are precision machines, dust can cause various problems, mixing with oil and grease during maintenance work and entering openings to foul contact points.

When connecting pipes and so on, the openings are usually plugged or capped to prevent the entry of dust particles, but, when darkness complicates the situation, it is difficult to determine whether or not dust has got in. Also, because ground tests are performed at engine rotation speeds that differ from the rotation speeds during flight, it is difficult to discover dust contamination.

In one incident (occurrence), the tip of an antidust plug broke off and remained in the pipe when a maintenance worker was attaching an oil pipe to the No. 2 gearbox (a case holding the gears that transmit the power of the engine). It was dark and the worker failed to notice the situation. He connected the pipe with the piece of plug still inside. During the leak tests on the ground, the engine was turned over by using the starter motor (driven by air pressure, and so the engine does not reach a high rotation speed), and therefore the oil pressure was not fully attained and there were no leaks. Once in flight, however, the oil pressure immediately became high and oil came out. The engine stopped and the plane turned back to the airport.

The degree of danger was 'considerable'.

In another case, an incident (occurrence) that had a 'considerable' degree of danger occurred when the Fuel Control Unit (FCU) of an engine was replaced. A non-specialist assisting maintenance worker dropped a seal and got a small amount of dust on it when installing a fuel pipe. He was supposed to wash the seal with fuel, but instead he wiped the seal with a rag without noticing that there was dust on the rag. As a result, there was dust on the seal when it was installed. No problems or abnormalities appeared in the leak test or when the engine was test run, so it passed inspection. However, a fuel leak from the engine was discovered a month later during a predeparture inspection. The degree of danger was 'considerable'.

4.15 Incidents caused by deterioration over time

Deterioration over time is a problem that is related to the ageing of equipment and refers to the reduction in the strength of structural elements of the airframe or problems in performance. Typical examples are corrosion from the effects of sulphur oxides or salt in the air and metal fatigue due to load and repeated vibration in flight.

The occurrence of deterioration over time is not related to maintenance work, but there are times when problems arise with the passage of time or the vibration of flight because of tension (tensile or compressive stress) on parts owing to excessive application of force or overtightening, or owing to adherence of human sweat during maintenance work. The discovery of deterioration in periodic inspections, particularly in the initial inspection of C maintenance, and the taking of appropriate measures are important tasks of maintenance.

The most representative accident of those in which deterioration over time was a main factor is the crash of a Boeing 707 cargo plane just before landing at Lusaka, Zambia, on 14 May 1977.[4] After use as a passenger plane in the United States, the plane that crashed had been purchased as a used aircraft by the Dunn Air Corporation, UK and then operated as a cargo plane. The accident happened when the horizontal stabilizer spar (a beam-like support for the stabilizer) collapsed immediately after 'full flaps', when the flaps are lowered to their maximum down position. The plane crashed in a nearly vertical descent from an altitude of about 1500 ft.

Examination of the horizontal stabilizer and elevator wreckage, which was found about 300 ft from the crash site in the direction from which the plane was flying, revealed that cracks had formed owing to fatigue from the 7000 flights over the 14 years that the plane had been in service after manufacture and had steadily increased since then. The photographs of that wreckage clearly show the conchoidal figure caused by metal fatigue, and have been presented in various places as typical examples.

One case in which deterioration over time was the only background factor is the following incident (discovery) for which the degree of danger was 'considerable'.

In the conditioning inspection that was done at the time of the previous C maintenance, cracking around the fill hole of the oil tank of the Auxiliary Power Unit (APU) was difficult to see because of the cap on the oil fill hole of the tank, and so the cracking was not discovered. In the subsequent T maintenance, however, it was discovered that the cap had come off completely owing to cracking from deterioration over time, and that oil was leaking.

At the shift change, the maintenance worker had already completed the external inspection and had confirmed the readiness for departure. All that was left was to hand the aircraft over to the flight crew (operation side). The reporter of the incident decided, 'Once more, just to make sure', and thus discovered that the area under the APU was dirtied by oil.

This kind of soiling is also something that can result from maintenance work, and is not necessarily the result of oil leaking from the APU. While it is possible to overlook this kind of thing in an easy-going manner, this maintenance worker went into the cockpit to check the oil gauge. Although the oil gauge showed that three quarts (about 2.8 l) of oil remained, which was within the specified range, he 'felt that, as the plane had undergone maintenance at Haneda Airport, the oil should have been topped off'. He then inspected the APU oil tank and discovered that the cap of the fill port had come off.

The prevention of an accident by the conscientious and scrupulous checking by a maintenance worker of an aircraft that had already passed inspection is worthy of praise.

4.16 Great difference between 'occurrence' and 'discovery'

Inspection and test/check investigation are also important elements of aircraft maintenance, in addition to the maintenance work itself. Because of the special nature of this work, incident reports that involve maintenance fall into two types: occurrence and discovery. This characteristic of incident reports that are related to maintenance was mentioned earlier.

Here, we take an overall look at the results of analysis for occurrence and discovery to see if the same results are obtained. This comparison, in fact, has important significance. The reason is that, if the results are the same for the two, then it would mean that the truth of the matter can be determined by investigation after the fact by someone other than the person directly involved; if the results are different, it would mean that after-the-fact investigation by another person is limited in effectiveness.

Firstly, let us look at the simple statistical results in Tables 4 and 5. Naturally, for both occurrence and discovery, individual incidents were analysed by the MAIR method described in Chapter 1. The results for occurrence and discovery differed in the proportions of the respective error modes and the background factors.

Concerning the error mode, for example, for occurrence incidents, recklessness or forcefulness and overdoing/overtightening in the action or instruction phase appear twice as often as for discovery incidents. Unreasonableness or high-handedness and excessive thoughtfulness or thinking too much in the decision phase also appear more for occurrence incidents, but incorrect or inappropriate action or instruction appears more for discovery incidents. Concerning background factors, there are 26 factors associated with occurrence incidents and 21

Table 4 Factors for quantification method III, aircraft maintenance 'occurrence' incidents involving human factors

Total number of incidents: 198

Total number of events: 316

		Categorical No.	Category	Frequency
Time of occurrence		1	Dawn / early morning	17
		2	Day	36
		3	Dusk	9
		4	Night / moonlight	80
Personal factors	Experience	5	Abundant experience / familiarity	16
		6	Insufficient experience / unfamiliarity	31
		7	Insufficient knowledge	31
	Physicals factor	8	Lack of sleep	7
		9	Fatigue	7
	Psychological factors	10	Impatience / fluster	26
		11	Inattentiveness (state of reaction after stress) / tedium	10
Team c-oordination		12	Excessive reliance / interdependence / excessive familiarity among team members	60
		13	Excessive concern / misplaced deference	5
		14	Insufficient communication	22
Parts / materials		15	Unavailable parts	8
		16	Unclear or absent label	6
Company policies		17	Keep schedule (event)	9
		18	Subcontracting / outside order	10
		19	Customary practice	5
Working environment		20	Good working environment	47
		21	Insecure footing	25
		22	Darkness (including inside tank)	59
		23	Tight area (including inside tank)	20
		24	Insufficient equipment or tools*	14
Special circumstances		25	Shift change / work interruption	36
		26	Time constraints	48
		27	Non-specialist	14
		28	Sudden occurrence	7
		29	Heavy workload / complicated or parallel work[†]	23
		30	Deterioration over time	12
Phase I (information source)		31	Discrepancy[‡]	9
		32	Modification order[§]	16
		33	Routine work card**	7
		34	Oral instructions	19
		35	Word of mouth	10
		36	Aircraft Maintenance Manual (AMM), etc.	16

	Categorical No.	Category	Frequency
	37	Target of work	217
	38	Work platforms	5
	39	Parts	11
Performer	40	Maintenance worker	202
	41	Other maintenance personnel (performer)	28
	42	Inspector	74
	43	Subcontractor	6
Phase II (information receiving)	44	Appropriate (information receiving)	121
	45	Did not see / hear	42
	46	Failure to see / hear / notice	66
	47	Saw incorrectly / heard incorrectly / mistaken impression	6
	48	Excessive focus	11
	49	Preconception	41
	50	Could not see / could not see clearly / could not hear / could not hear clearly / could not reach / hard to reach	22
Phase III (decision)	51	Appropriate (decision)	18
	52	Forgot (decision)	52
	53	Easy-going / assumption	108
	54	Wrong conviction / hasty conclusion / one's own interpretation	74
	55	Vacillation	6
	56	Unreasonableness / high-handedness / excessive thoughtfulness / thinking too much	25
	57	Insufficient consideration	6
	58	Difficult (decision) / inability to decide	19
Phase III (action / instruction)	59	Forgot (action / instruction)	58
	60	Aimlessness / sat idle / instructions not received clearly	21
	61	Incorrect action / inappropriate action	130
	62	Loss of balance / instability / lack of effective teamwork / ambiguous or vague instruction	8
	63	Recklessness / forceful action / overdoing / overtightening	55
	64	Insufficient / incomplete	32
	65	Difficult (action / instruction) / inability to work	7
Incident critical phase	66	Information source	26
	67	Information receiving	95
	68	Decision	82
	69	Action / instruction	113
Reasons for recovery	70	Notice own error	15

	Categorical No.	Category	Frequency
	71	Other maintenance personnel (recovery)	49
	72	Test	41
	73	Inspection	5
	74	Pointing out by crew	115
	75	Occurrence of consequence	81
	76	Good fortune	5
Degree of danger	77	Virtually none	103
	78	Fairly little	104
	79	Considerable	86
	80	Serious	23

The categories referred to in quantification method III; that is, the factors in this book.

* Insufficient equipment or tools: this refers to a wide range of equipment and tools, from screwdrivers to cranes for suspending engine and measuring devices used for instrument calibrations.

† Complicated or parallel work: this could refer to cases in which multiple workers (groups) are working simultaneously in the same space or in nearby spaces, or when separate work is being done on the same system at the same time, or a combination of both.

‡ Discrepancy: a card on which the maintenance worker who discovered a given condition or malfunction records the details accordingly, and records the work that was required.

§ Modification order: a document giving instructions for changes in specifications intended to improve airworthiness, customer service, handling efficiency, etc. The name of this order differs from one company to the next.

** Routine work card: a work instruction document that records the maintenance work that should be done on a regular basis.

factors associated with discovery incidents. Of those, 17 factors are common to the two types of incident and the others are specific to each type of incident.

For discovery, there are very few personal factors and there are no physical factors at all. Psychological factors are also very few. The special circumstance factors of sudden occurrences, unclear or absent labelling, and insecure footing appear only for occurrence incidents, and the defective parts, dust, and painting and polishing factors appear only for discovery incidents. Also, even the background factors that are common to the two types of incident differ considerably in their proportions with respect to the total number of factors for the two types of incident. The occurrence-type incidents had a higher proportion of shift change or work interruption and time constraints to the total number of factors, while the discovery-type incidents had a higher proportion of the deterioration over time factor.

These results show that there are limits to the description of the errors and the circumstances at the time that the work was done when the reporting person is not the person who performed the work.

Table 5 Factors for quantification method III, aircraft maintenance 'discovery' incidents involving human factors

Total number of incidents: 153

Total number of events: 250

	Categorical No.	Category	Frequency
Personal factors			
Experience / knowledge	1	Insufficient experience / unfamiliarity	25
	2	Insufficient knowledge	20
Team co-ordination	3	Excessive reliance / interdependence	6
	4	Insufficient communication	14
Parts / materials	5	Unavailable parts	5
	6	Defective parts	19
Company policies	7	Keep schedule (event)	5
	8	Subcontracting / outside order	22
	9	Customary practice	5
Working environment	10	Good working environment	11
	11	Darkness	16
	12	Tight area	19
	13	Dust	7
	14	Painting / polishing	19
	15	Insufficient equipment / tools*	8
Special circumstances	16	Shift change / work interruption	12
	17	Time constraints	17
	18	Non-specialist	16
	19	Heavy workload / complicated or parallel work[†]	16
	20	Manufacturing error	53
	21	Deterioration over time	44
Phase I (information source)	22	Modification order[‡]	7
	23	Routine work card[§]	7
	24	Oral instructions	6
	25	Aircraft Maintenance Manual (AMM), etc.	11
	26	Target of work	150
	27	Tools	9
	28	Parts	37
Performer	29	Maintenance worker	99
	30	Other maintenance personnel (performer)	11
	31	Inspector	75
	32	Manufacturer	39
	33	Subcontractor	17
Phase II (information receiving)	34	Appropriate (information receiving)	80
	35	Did not see / hear	27
	36	Failure to see / hear / notice	53
	37	Excessive focus	8
	38	Preconception	24

	Categorical No.	Category	Frequency
	39	Could not see / could not see clearly / could not hear / could not hear clearly / could not reach / hard to reach	32
Phase III (decision)	40	Appropriate (decision)	8
	41	Forgot (decision)	35
	42	Easy-going / assumption	75
	43	Wrong conviction / hasty conclusion / one's own interpretation	38
	44	Unreasonableness / high-handedness / excessive thoughtfulness / thinking too much	14
	45	Insufficient consideration	7
	46	Difficult (decision) / inability to decide	48
Phase IV (action / instruction)	47	Forgot (action / instruction)	42
	48	Aimlessness / sat idle	22
	49	Incorrect action / inappropriate action	122
	50	Recklessness / forceful action / overdoing / overtightening	23
	51	Insufficient / incomplete	21
	52	Difficult / inability to work (action / instruction)	12
Incident critical phase	53	Information source	26
	54	Information receiving	76
	55	Decision	74
	56	Action / instruction	72
Reasons for recovery	57	Other maintenance personnel (recovery)	90
	58	Test	23
	59	Inspection	6
	60	Pointing out by crew	90
	61	Occurrence of consequence	30
	62	Good fortune	6
Degree of danger	63	Virtually none	61
	64	Fairly little	66
	65	Considerable	79
	66	Serious	44

The categories referred to in quantification method III; that is, the factors in this book.

* Insufficient equipment / tools: this refers to a wide range of equipment and tools, from screwdrivers to cranes for suspending engine and measuring devices used for instrument calibrations.

† Complicated or parallel work: this could refer to cases in which multiple workers (groups) are working simultaneously in the same space or in nearby spaces, or when separate work is being done on the same system at the same time, or a combination of both.

‡ Modification order: a document giving instructions for changes in specifications intended to improve airworthiness, customer service, handling efficiency, etc. The name of this order differs from one company to the next.

§ Routine work card: a work instruction document that records the maintenance work that should be done on a regular basis.

4.17 Actual situation of latent danger factors versus general recognition

To see what the correlations among the various factors are now, we applied quantification method III to the respective results for occurrence and discovery incidents (see Tables 4 and 5). The results are shown in the following.

Looking from the front of the three-dimensional data display at the X and Y axes, for which the correlations appear most strongly, we see that a systematic factor group and a personal factor group rise opposite each other on the left and right, representing two major characteristics as shown in Fig. 22.

The background factors that constitute the systematic factor group are insufficient equipment or tools, unavailability of parts, insecure footing, deterioration over time, non-specialist, insufficient experience or unfamiliarity, tight area, misplaced deference or excessive concern, unclear or absent label, subcontracting, and keep schedule. The error modes were 'could not see clearly, could not reach sufficiently or hard to reach' in the information receiving phase; 'easy-goingness or assumption', 'vacillation', 'unreasonableness or excessive thoughtfulness', or 'difficult decision/inability to decide' in the decision phase; and 'sat idle/ aimlessness', 'loss of balance/stability', 'recklessness, forceful action or overdoing', 'insufficient or incomplete work', and 'difficult action or instruction' or 'inability to act or work' in the action or instruction phase.

The background factors that constitute the personal factor group, on the other hand, are fatigue, lack of sleep, impatience or fluster, inattentiveness (state of reaction after stress), shift change or work interruption, and heavy workload. The error modes are 'did not see or did not hear', 'excessive focus', 'saw incorrectly or heard incorrectly' in the information receiving phase, and 'forgot' in the decision phase and the action or instruction phase.

The following factors are located between the systematic factor group and the personal factor group, which are different in nature, and are related to both factor groups (referred to as bipolar related factors): insufficient communication, sudden occurrence, darkness, customary practice, time constraints, word of mouth, heavy workload, insufficient knowledge, excessive reliance or interdependency, and abundant experience or familiarity. Customary practice, shift change or work interruption, and heavy workload are at the same time personal factors and factors that are related to the two groups.

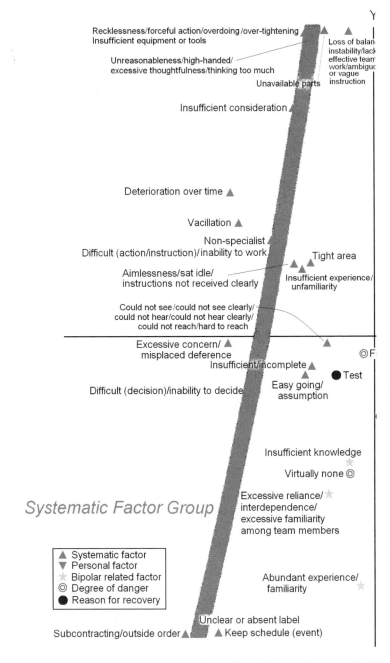

Fig. 22 Two major characteristics and bipolar related factor 'occurrence' incidents. Aircraft maintenance personnel and latent danger in aircraft (see colour plate section)

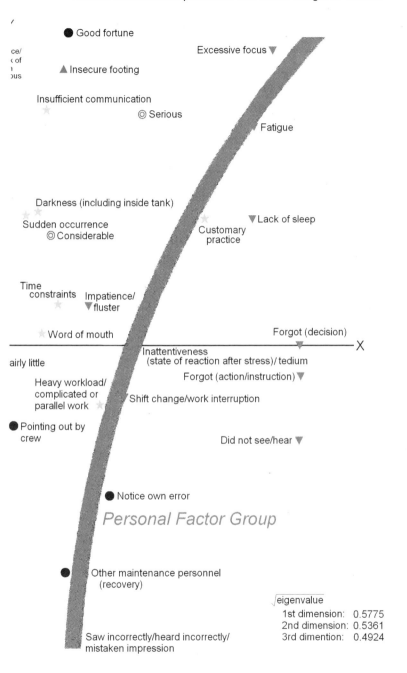

● Good fortune

Excessive focus ▼

ce/
< of
า
ɔus

▲ Insecure footing

Insufficient communication

◎ Serious

Fatigue

Darkness (including inside tank)

Sudden occurrence
◎ Considerable

Customary
practice

▼ Lack of sleep

Time
constraints   Impatience/
▼ fluster

Word of mouth

Forgot (decision)
▼
—X

Inattentiveness
(state of reaction after stress)/ tedium

airly little

Forgot (action/instruction) ▼

Heavy workload/
complicated or
parallel work

Shift change/work interruption

● Pointing out by
crew

Did not see/hear ▼

● Notice own error

Personal Factor Group

● Other maintenance personnel
(recovery)

√eigenvalue
1st dimension: 0.5775
2nd dimension: 0.5361
3rd dimention: 0.4924

Saw incorrectly/heard incorrectly/
mistaken impression

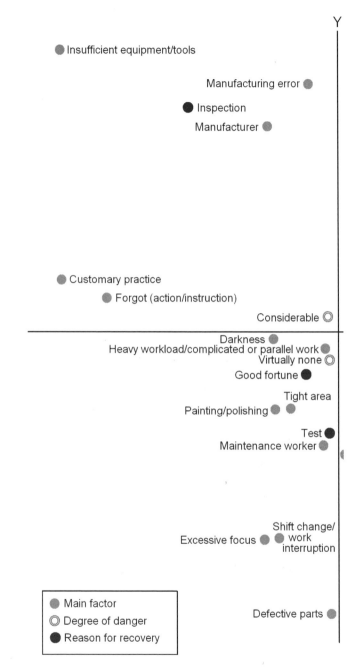

Fig. 23 Distribution of main factors for 'discovery' incidents. Aircraft maintenance personnel and latent danger in aircraft (see colour plate section)

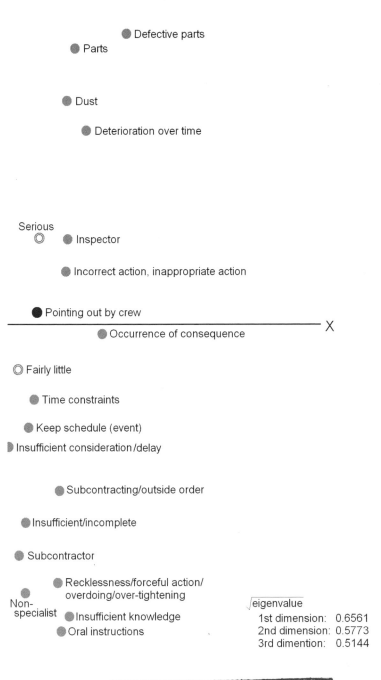

Defective parts

Parts

Dust

Deterioration over time

Serious

Inspector

Incorrect action, inappropriate action

Pointing out by crew

Occurrence of consequence   X

Fairly little

Time constraints

Keep schedule (event)

Insufficient consideration/delay

Subcontracting/outside order

Insufficient/incomplete

Subcontractor

Recklessness/forceful action/
overdoing/over-tightening

Non-
specialist   Insufficient knowledge

Oral instructions

√eigenvalue
1st dimension: 0.6561
2nd dimension: 0.5773
3rd dimention: 0.5144

This fact suggests that, when these bipolar related factors are involved, a link can be made between the two factor groups. Accordingly, to lower the rate at which unsafe events occur in the field of maintenance, it is natural that improvements be made for the factors that are in the two factor groups and are associated with a high degree of danger. It is also very important, however, that measures be taken for the bipolar factors (see Fig. 22) so as to reduce the possibility of linking the two factor groups.

Next, looking at the above-mentioned dimension for which the correlations are strongest for the discovery incidents, we do not see the two opposing factor groups that we saw for the occurrence incidents. Rather, except for a very few factors, the factors are all located in close proximity to the centre, a strikingly different pattern from that of the occurrence incidents (see Fig. 23).

The four factors that represent degree of danger, too, are found near the centre and not in order, whereas, for the occurrence incidents, they are distributed at appropriate intervals and in order of degree of danger. Also, grouping the factors that have strong correlations, we see that there are 13 clusters for occurrence and nine for discovery. Aside from the 'forgot' cluster, the two incident types do not have any clusters in common for which the main constituent factors are the same.

If we consider why the analytical results for occurrence and discovery incidents differ in this way, we notice that there are differences in the incident reports themselves, which are the foundation of the analysis. Both occurrence and discovery are the same in terms of the specific descriptions of the things that occurred, and there is no difference in terms of quality, but, on the point of the process through which the problem occurred, the two differ.

In the reports of occurrences, the error made by the person writing the report and the circumstances at the time the work was done, which include the background factors, are described. The actual facts are described in detail and accurately so that what kind of error is made and why, that is to say, the correlations between the error modes and the background factors – the process of the incident occurrence – can be analysed. Accordingly, it can be said that the analysis results for occurrence incidents show the actual state of the matter for the latent danger factors at the time of the research in the field of aircraft maintenance.

In contrast, the discovery reports are written by someone other than the person involved in the error, so there is naturally little information available regarding the circumstances of the persons involved in the errors, even aside from the description of the error itself. Beyond this, however, there is little information on the intertwining of the background factors that are related to the human factors of the person who performed the work, because a certain amount of time has passed from the time the problem occurred to the time it was discovered. Accordingly, one must wonder if inferences by the discoverer, who writes the report, are not inserted in the places where information is insufficient because of the limited information that is available.

With this kind of thinking, it can be said that what is reflected in the analysis results for discovery incidents is a general understanding of the field of aircraft maintenance rather than the actual circumstances.

4.18 Limits of *ex post facto* investigations and the need for the IRAS approach

Here, we will try to discuss further, with respect to the degrees of danger of the various factors, whether or not there is a difference between occurrence and discovery, that is to say, between the actual situation and the general understanding, which is extremely important from the perspective of maintaining safety and preventing accidents.

Firstly, we compared the analysis results of quantification method III applied to the 60 factors that are common to the two types of incident, occurrence and discovery (see Table 6). Those results showed that no more than 22 of those factors, less than half, had the same evaluation for both types of incident (see Table 7). What must be noted here is that the factors are considered to be associated with a low for discovery, which is to say the general understanding, even though the degree of danger is high for occurrence, or the actual situation.

The factors that are relevant here are 'unreasonableness, high-handedness, excessive thoughtfulness or thinking too much', and 'insufficient consideration' in the decision phase, 'recklessness, forceful-ness, overdoing or overtightening' in the action or instruction phase, and 'unavailable parts'.

It can be expected that improvements and countermeasures will be taken for the factors that are associated with a high degree of danger for both types of incident and those that are associated with a high degree of danger in the general understanding case, but are associated with a low degree of danger in the actual situation case. Unless IRAS is implemented, however, the factors that are associated with a high degree of danger in the actual situation but with a low degree of danger in the general understanding will not be recognized as having a high degree of danger, and thus studies of countermeasures for them may be neglected.

Table 6 Comparison of degree of danger (both 'occurrence' and 'discovery' use second root categorical values), sixty common 'Occurrence' and 'Discovery' factors in aircraft maintenance

Occurrence			Discovery		
Category No.	Category	Categorical value	Category No.	Category	Categorical value
38	Unreasonableness / high-handedness / excessive thoughtfulness / thinking too much	5.641	47	Information source	4.741
56	Good fortune	5.033	40	Difficult (decision) / inability to work	4.34
31	Excessive focus	4.989	53	Inspection	3.761
5	Unavailable parts	4.601	33	Could not see / could not see clearly / could not hear / could not hear clearly / could not reach / hard to reach	3.493
44	Recklessness / forceful action / overdoing	4.512	12	Insufficient equipment / tools	3.039
12	Insufficient equipment or tools	3.738	23	Parts	2.977
55	Occurrence of consequence	3.611	26	Inspector	2.603
4	Insufficient communication	3.526	17	Deterioration over time	2.471
60	Serious	3.322	16	Heavy workload / complicated or parallel work	2.162
39	Insufficient consideration, delay	2.823	29	Did not see / hear	1.713
8	Customary practice	2.64	32	Preconception	1.508
49	Decision	2.19	43	Incorrect action / inappropriate action	1.427
34	Appropriate (decision)	1.803	8	Customary practice	1.36
17	Deterioration over time	1.746	35	Forgot (decision)	1.296
59	Considerable	1.47	46	Difficult / inability to work (action / instruction)	1.118
10	Darkness (including inside tank)	1.29	60	Serious	1.115
46	Difficult / inability to work (action / instruction)	1.221	56	Good fortune	1.066
28	Appropriate (information receiving)	0.923	6	Keep schedule (event)	0.907
1	Insufficient experience / unfamiliarity	0.877	41	Forgot (action / instruction)	0.71
11	Tight area (including inside tank)	0.866	54	Pointing out by crew	0.586
22	Target of work	0.864	59	Considerable	0.394
25	Other maintenance personnel (performer)	0.766	48	Information receiving	0.335
24	Maintenance worker	0.76	55	Occurrence of consequence	0.18
14	Time constraints	0.623	37	Wrong conviction / hasty conclusion / one's own interpretation	0.082
27	Subcontractor	0.467	3	Excessive reliance / interdependence / excessive familiarity among team members	0.05
15	Non-specialist	0.429	49	Decision	-0.268
50	Action / instruction	0.411	14	Time constraints	-0.272
35	Forgot (decision)	0.3	51	Other maintenance personnel (recovery)	-0.292

About 2●
(14 cate●

	Occurrence			Discovery	
Category No.	Category	Categorical value	Category No.	Category	Categorical value
42	Aimlessness / sat idle / instructions not received clearly	0.237	19	Routine work card	-0.368
41	Forgot (action / instruction)	-0.038	22	Target of work	-0.387
33	Could not see / could not see clearly / could not hear / could not hear clearly / could not reach / hard to reach	-0.12	18	Modification order	-0.451
52	Test	-0.296	58	Fairly little	-0.454
58	Fairly little	-0.378	10	Darkness	-0.47
16	Heavy workload / complicated or parallel work	-0.454	57	Virtually none	-0.589
29	Did not see / hear	-0.723	28	Appropriate (information receiving)	-0.667
36	Easy-going / assumption	-0.734	4	Insufficient communication	-1.081
45	Insufficient / incomplete	-0.836	36	Easy-going / assumption	-1.118
13	Shift change / work interruption	-0.877	7	Subcontracting / outside order	-1.372
30	Failure to see / hear / notice	-0.94	24	Maintenance worker	-1.406
54	Pointing out by crew	-1.088	30	Failure to see / hear / notice	-1.519
40	Difficult (decision) / inability to work	-1.164	50	Action / instruction	-1.636
32	Preconception	-1.229	25	Other maintenance personnel (performer)	-1.713
37	Wrong conviction / hasty conclusion / one's own interpretation	-1.486	13	Shift change / work interruption	-1.847
48	Information receiving	-1.494	9	Good working environment	-1.9
21	Aircraft Maintenance Manual (AMM), etc.	-1.505	21	Aircraft Maintenance Manual (AMM), etc.	-1.911
57	Virtually none	-1.651	11	Tight area	-2.017
2	Insufficient knowledge	-1.773	31	Excessive focus	-2.162
19	Routine work card	-1.811	42	Aimlessness / sat idle / instructions not received clearly	-2.248
43	Incorrect action / inappropriate action	-2.07	1	Insufficient experience / unfamiliarity	-2.352
20	Oral instructions	-2.117	52	Test	-2.413
26	Inspector	-2.221	27	Subcontractor	-2.648
3	Excessive reliance / interdependence / excessive familiarity among team members	-2.524	39	Insufficient consideration, delay	-2.733
9	Good working environment	-2.579	15	Non-specialist	-3.258
51	Other maintenance personnel (recovery)	-3.003	45	Insufficient / incomplete	-3.347
47	Information source	-3.474	5	Unavailable parts	-3.638
53	Inspection	-3.602	34	Appropriate (decision)	-4.41
23	Parts	-3.983	2	Insufficient knowledge	-4.646
6	Keep schedule (event)	-4.374	44	Recklessness / forceful action / overdoing	-4.784
18	Modification order	-4.668	38	Unreasonableness / high-handedness / excessive thoughtfulness / thinking too much	-4.986
7	Subcontracting / outside order	-4.727	20	Oral instructions	-5.553

About 60% (34 categories)

About 20% (12 categories)

Expressions for categories have been abbreviated.
The categorical value referred to in quantification method III; that is, the value computed for each of the factors in this book.

Table 7 Classification of degree of danger by groups, sixty common 'occurrence' and 'discovery' factors in aircraft maintenance

Occurrence	*Discovery* +	*Discovery* +−	*Discovery* −
+	(i) Insufficient equipment or tools Customary practice Deterioration over time	(iv) Good fortune Excessive focus Occurrence of consequence Insufficient communication Serious Decision	(viii) Unreasonableness / high-handedness / thinking too much Unavailable parts Recklessness / forceful action / overdoing Insufficient consideration (decision) Appropriate (decision)
+−	(v) Forgot (decision) Could not see / could not see clearly / could not reach / hard to reach Heavy workload / complicated or parallel work Did not see / hear (information receiving) Difficult (decision) Preconception	(ii) Considerable Darkness Difficulty / inability to work (action) Appropriate (information receiving) Tight area Target of work Other maintenance personnel (performer) Maintenance worker Time constraints Action / instruction Aimlessness / sat idle Forgot (action) Fairly little Easy-going / assumption Shift change / work interruption Failure to see / hear / notice Pointing out by crew Wrong conviction Information receiving Aircraft Maintenance Manual (AMM), etc. Virtually none Routine work card	(vi) Insufficient experience / unfamiliarity Subcontractor Non-specialist Test Insufficient / incomplete (action / instruction) Insufficient knowledge
−	(ix) Incorrect action / inappropriate action Inspector Information source Inspection (recovery) Parts	(vii) Excessive reliance / interdependence Subcontracting / outside order Good working environment Other maintenance personnel (recovery) Keep schedule (event) Modification order	(iii) Oral instructions

Expressions for categories have been abbreviated.

Also, there are five factors that appear only in the occurrence incidents (loss of balance/stability, insecure footing, work platform, fatigue, and dusk) and there are two factors that appear only in the discovery incidents (painting/polishing and dust) (see Table 8).

In Section 4.16 we raised the question of whether or not the truth can be determined by investigations conducted after the fact by a person other than the person directly involved in the work as well as it can by description of the circumstances at the time by the person himself. In answer to this question, we must say that, according to the verification described above, it is clear that *ex post facto* investigation is limited.

As can be seen in the examples presented in Chapter 1, the process of occurrence of unsafe events for both incidents and accidents is an accidental, illogical, and irrational actual chain of danger factors. Rather than something that is necessarily inevitable, logical, and rational, it is most often far from something that can be considered from a common sense point of view. Accordingly, its special character is that ordinary rule-of-thumb inferences are not valid.

From the fact that objective validity is required in *ex post facto* investigation, unless clear evidence can be shown, the attempt to achieve a single logical consistency in the investigation results may lead to conclusions that differ from the actual facts.

The cause of the differences in the results of analysis for occurrence and discovery incidents can be considered the fact that, based on limited information, the logic of the discoverer has been incorporated into the results for discovery incidents where there is insufficient information. Expressed in a different way, in the world of logic there is no logic that recognizes the irrational as irrational, so the conclusions may be made by stringing together part of the factual data obtained from investigation results and marginal opinions formed from experiments or tests premised on assumed conditions with the aim of logical conformity.

The incident discovery and the investigation conducted after an accident differ greatly in terms of the scope and scale of investigation. They are essentially the same, however, on the point of being *ex post facto* investigations made by persons other than the persons directly involved and thus requiring objective validity.

In the case of a serious accident, it is extremely difficult to determine the true process of the occurrence of the accident that involves human factors because the persons directly involved may be dead, or, if still alive, exercising their right to be silent, and because of damage to the aircraft equipment. Accordingly, although improvement measures that are derived from the results of accident investigations, too, play a role in maintaining safety in a broad sense, they are not necessarily effective in preventing exactly the same type of accident from happening again.

We can say that the differences between occurrence and discovery concerning maintenance, which are seen in the results of comprehensive analysis by IRAS, and the similarity of the two within a certain range, including the case of accidents, show that there are limits to the understanding of the actual situation by means of *ex post facto* investigations conducted by persons other than those directly involved. These factors also show how difficult it is to point out the differences between the results of *ex post facto* investigations and the actual situation.

First of all, a comprehensive understanding of the actual situation concerning the latent danger factors of the field is a major prerequisite for the prevention of accidents and the preservation of safety, as has been described repeatedly in this book. Towards that end, reports of the true facts of the situation by the persons who themselves participated in the occurrence of human-factor incidents are indispensable. Unless based on that kind of information, a comprehensive, multidimensional understanding of the true factual information cannot be obtained and appropriate safety countermeasures cannot be taken.

The differences in the analysis results for occurrence and discovery are sufficient to make us appreciate all the more the necessity of IRAS.

Table 8 Degree of danger in each 'occurrence' and 'discovery' factor in aircraft maintenance ('occurrence' uses second root and 'discovery' uses third root categorical values)

	Occurrence (80)				Discovery (66)			
Category No.	Proper categories[1]	Category	Categorical value	Category No.	Proper categories[2]	Category	Categorical value
62	*	Loss of balance or stability / lack of effective teamwork / ambiguous or vague instruction	5.299	62		Good fortune	5.828
56		Unreasonableness / high-handedness / excessive thoughtfulness / thinking too much	4.77	7		Keep schedule (event)	5.553
3	*	Dusk	4.461	19		Heavy workload / complicated or parallel work	4.87
48		Excessive focus	4.371	3	*	Excessive reliance / interdependence	4.524
24		Insufficient equipment or tools	4.347	14		Painting / polishing	3.158
76		Good fortune	4.004	31		Inspector	3.031
63		Recklessness / forceful action / overdoing / overtightening	3.947	39		Could not see / could not see clearly / could not hear / could not hear clearly / could not reach / hard to reach	2.862
15		Unavailable parts	3.692	17		Time constraints	2.782
75		Occurrence of consequence	3.685	37		Excessive focus	2.743
38	*	Work platforms	3.547	46		Difficult (decision) / inability to decide	2.612
21	*	Insecure footing	3.325	66		Serious	2.536
14		Insufficient communication	2.914	16		Shift change / work interruption	2.333
80		Serious	2.889	4		Insufficient communication	2.161
9	*	Fatigue	2.787	54		Information receiving	1.479
57		Insufficient consideration	2.78	41		Forgot (decision)	1.476
51		Appropriate (decision)	2.388	9	*	Customary practice	1.464
68		Decision	1.992	5		Unavailable parts	1.421
30	*	Deterioration over time	1.803	26		Target of work	1.236
1		Dawn / early morning	1.728	13		Dust	1.133
22	*	Darkness (including inside tank)	1.669	55	*	Decision	0.907
28		Sudden occurrence	1.638	35		Did not see / hear	0.872
19	*	Customary practice	1.629	42		Easy-going / assumption	0.859
8		Lack of sleep	1.587	59		Inspection	0.766
79		Considerable	1.434	60		Pointing out by crew	0.725
20% (16 categories)							20% (13 categories)

Occurrence (80)				Discovery (66)			
Category No.	Proper categories	Category	Categorical value	Category No.	Proper categories	Category	Categorical value
55	*	Vacillation	1.341	47		Forgot (action / instruction)	0.694
44		Appropriate (information receiving)	1.168	49		Incorrect action, inappropriate action	0.667
27		Non-specialist	1.126	48		Aimlessness / sat idle	0.648
65	*	Difficult (action / instruction) / inability to work	1.06	52		Difficult / inability to work (action / instruction)	0.559
4	*	Night / moonlight	0.903	1		Insufficient experience / unfamiliarity	0.377
6		Insufficient experience / unfamiliarity	0.885	36		Failure to see / hear / notice	0.339
23		Tight area (including inside tank)	0.879	61		Occurrence of consequence	0.337
60		Aimlessness / sat idle / instructions not received clearly	0.816	30		Other maintenance personnel (performer)	0.096
37		Target of work	0.804	65		Considerable	-0.097
41		Other maintenance personnel (performer)	0.796	29		Maintenance worker	-0.368
69		Action / instruction	0.712	64		Fairly little	-0.464
40		Maintenance worker	0.618	8		Subcontracting / outside order	-0.498
43		Subcontractor	0.55	21		Deterioration over time	-0.594
26		Time constraints	0.53	34		Appropriate (information receiving)	-0.651
10	*	Impatience / fluster	0.525	45		Insufficient consideration	-0.858
35	*	Word of mouth	0.083	18		Non-specialist	-0.967
11	*	Inattentiveness (state of reaction after stress) / tedium	0.037	57		Other maintenance personnel (recovery)	-1.023
52		Forgot (decision)	-0.047	11		Darkness	-1.075
50		Could not see / could not see clearly / could not hear / could not hear clearly / could not reach / hard to reach	-0.099	25		Aircraft Maintenance Manual (AMM), etc.	-1.087
13	*	Excessive concern / misplaced deference	-0.137	6	*	Defective parts	-1.153
78		Fairly little	-0.237	53		Information source	-1.44
2	*	Day	-0.27	28		Parts	-1.454
59		Forgot (action / instruction)	-0.354	23		Routine work card	-1.47
53		Easy-going / assumption	-0.436	51		Insufficient / incomplete (action / instruction)	-1.474
64		Insufficient / incomplete (action / instruction)	-0.446	63		Virtually none	-1.518
72		Test	-0.538	38		Preconception	-1.723
25		Shift change / work interruption	-0.681	43		Wrong conviction / hasty conclusion / one's own interpretation	-1.815
29		Heavy workload / complicated or parallel work	-0.712	33		Subcontractor	-1.89

60% (39 categories)

Occurrence (80)

Category No.	Proper categories[1]	Category	Categorical value
46		Failure to see / hear / notice	-0.859
74		Pointing out by crew	-0.903
45		Did not see / hear	-1.076
34		Oral instructions	-1.224
49	*	Preconception	-1.254
7		Insufficient knowledge	-1.637
36		Aircraft Maintenance Manual (AMM), etc.	-1.691
77		Virtually none	-1.715
54		Wrong conviction / hasty conclusion / one's own interpretation	-1.725
67		Information receiving	-1.768
33		Routine work card	-1.794

60% (48 categories)

Category No.	Proper categories[1]	Category	Categorical value
42		Inspector	-1.978
12		Excessive reliance / interdependence / excessive familiarity among team members	-2.047
70	*	Notice own error	-2.064
61	*	Incorrect action / inappropriate action	-2.086
31		Discrepancy	-2.469
20		Good working environment	-2.578
71		Other maintenance personnel (recovery)	-2.895
5	*	Abundant experience / familiarity	-3.094
66		Information source	-3.463
39		Parts	-3.916
17		Keep schedule (event)	-4.225
16	*	Unclear or absent label	-4.275
18		Subcontracting / outside order	-4.319
73		Inspection	-4.517
32		Modification order	-4.919
47	*	Saw / heard incorrectly / mistaken impression	-5.65

20% (16 categories)

Discovery (66)

Category No.	Proper-categories[1]	Category	Categorical value
22		Modification order	-2.188
56		Action / instruction	-2.233
58		Test	-2.342
27	*	Tools	-2.426
10		Good working environment	-2.548
12		Tight area	-2.597
20	*	Manufacturing error	-3.778
2		Insufficient knowledge	-3.988
50		Recklessness / forceful action / overdoing / overtightening	-4.71
32	*	Manufacturer	-4.782
40		Appropriate (decision)	-4.811
44		Unreasonableness / high-handedness / excessive thoughtfulness / thinking too much	-4.95
24		Oral instructions	-5.682

20% (14 categories)

Expressions for categories have been abbreviated.
The categories referred to in quantification method III; that is, the factors in this book.
[1] Factors seen only in 'occurrence' (factor)
[2] Factors seen only in 'discovery' factor

Epilogue

Towards the establishment of IRAS

E.1 What have we learned from drug disasters?

Maintaining safety for both human beings and the Earth is without doubt one of the most important issues for the twenty-first century, when even further progress and development are expected in science and technology.

That there are events that necessarily occur in advance of disasters and major incidents, events that we should call danger signs, is not limited to aircraft accidents and accidents that involve other such large and complex systems, but extends to broad social incidents as well. Taking measures to deal with such events, rather than taking them lightly, is the best method for preventing accidents and incidents before they occur.

Chapter 1 introduced several examples of accidents, beginning with the nuclear power plant accident at Chernobyl, but time prevented the mentioning of examples in which such events that signal danger were not dealt with appropriately and so the situation reached grave proportions.

Lately in Japan, there was the problem of an AIDS outbreak that was caused by medical treatment, which called medical ethics into question.

Previously, in 1983, it had been suggested in *Disease Weekly*, a publication of the Center for Disease Control in the United States, that unheated blood serum concentrate posed a danger[1] and a heat treatment formulation was endorsed in 1983.[2] In spite of that, with no scientific studies of the matter having been done in Japan, physicians specializing in haemophilia, pharmaceutical companies (one of which had reported to the Ministry of Health and Welfare the recall of some imported blood products that may have been contaminated with the AIDS virus),[3] and the Ministry of Health and Welfare denied the danger[4] and continued to import virus-infected blood products from the

United States until 1985[5] and continued to use those products in medical treatment. Because of that, many haemophiliac patients contracted AIDS and died from it (by November 1997, 400 of these patients had died).

The medicine-induced AIDS problem occurred in spite of the fact that it had been preceded by various pharmaceutical disasters. In 1979, in order to protect against the occurrence and spread of dangers to health and sanitation through medicinal products, the Drugs, Cosmetics and Medical Instruments Act was revised to grant the Minister of Health and Welfare the authority to order a temporary halt to the sale of pharmaceuticals and the disposal or recall of defective medicines, and the authority to provide information to physicians by means of Doctor's Letters (reports of emergency safety information). In spite of that, no measures were ever taken.

Prior to the AIDS problem, there had been three pharmaceutical disasters, which respectively involved the chemical compounds thalidomide, chinoform, and chloroquine, and are known as the three great medical disasters of Japan.

Concerning the thalidomide problem, in November 1961, Dr Lentz made an announcement to the Institute of Pediatrics of West Germany, warning pharmaceutical companies and the government of the danger of the birth of deformed infants. Within 3 days of that, all of the drug had been recalled in West Germany. In Japan, pharmaceutical companies voluntarily stopped shipment of that drug in May 1962, 6 months after Dr. Lentz's announcement, and recall of the drug was begun in September of that year, 10 months after the announcement. Up to that time, no measure was taken by either the Ministry of Health and Welfare or medical institutes. In that time, the rate of thalidomide-induced birth defects is said to have increased sharply.[6]

Moreover, chinoform, the substance that caused the SMON disease (subacute myelo-opticoneuropathy), which affected over 10 000 persons in Japan, becoming the world's largest medical disaster, had originally caught attention as a specific drug for amoebic dysentery, but it quickly became used as a general therapeutic drug for diarrhoea. In Japan, in particular, it was used in long-term treatment for intestinal disorders.[7] Because of that, many people fell victim to the SMON disease.

It is probably not possible to avoid side effects in medicines. Medicinal drugs must be administered according to the patient's condition and in appropriate amounts considering both the side effects and the efficacy of the drug. No matter how effective a drug might be, it is a matter of course that its overuse should be avoided. The efficacy of a drug includes side effects, so should not objective studies have been conducted? In addition, should not fast and appropriate measures have been taken in response to the report of side effects?[8]

For both the thalidomide problem and the SMON disease issue, the Ministry of Health and Welfare did not even set up an investigative team.

The SMON disease was not prevalent in Sweden because a Swedish paediatric neurologist by the name of Hanson announced in 1966 an outbreak of visual impairment and warned Ciba-Geigy, the manufacturer, and medical institutes. The government officials responsible for pharmaceuticals were responsive to the information and took countermeasures. In Japan, the notification concerning chinoform was issued by the head of the Pharmaceuticals Affairs Bureau of the Ministry of Health and Welfare in September 1970, and sales of the drug were stopped one and a half years after regulations were imposed in Sweden.

Chloroquine, developed as a drug for the treatment of malaria, was held also to be effective for kidney ailments and epilepsy (actually, there are also indications that it had no such efficacy) and administered for such treatment. With long-term use in high volumes, however, the side effect of damage to the eyes, such as loss of eyesight owing to retinopathy, arose.[9] Sales of imports in Japan began in 1955, but, in 1959, detailed case reports of chloroquine retinopathy were published in a medical journal in the United Kingdom.[10] In spite of that, production of chloroquine was begun in Japan in 1961.[11]

In the autumn of 1962, cases of chloroquine retinopathy were reported in Japan as well.[12] Three years after that, in 1965, the head of the Ministry of Health and Welfare section that was responsible for these matters at that time obtained information on the harmful effects of chloroquine and, knowing the danger of the side effects, personally stopped using it.[13] The Ministry of Health and Welfare, however, did not issue a notification for caution in the use of the drug in the name of the head of the Pharmaceutical Affairs Bureau until 1969,[14] and the manufacturer did not halt production of the drug until 1974.[15] The drug was not cleared from the market until the following year, 1975.[16]

Test calculations show that, if the government had taken measures to prevent harm on the basis of the information about the side effects in 1965, 80 per cent of the patients could have been spared the disease.[17] Be that as it may, the number of persons afflicted with this disease increased while the government bureaus concerned failed to take appropriate measures without feeling even the slightest misgivings.

Concerning the issue of life-threatening side effects, even if the cause-and-effect relationship between drug administration and disease is not necessarily scientifically demonstrated, treating information that is received on side effects as incidents and taking measures to cope with them can be said to be a prudent attitude in terms of humanity.

E.2 'Incidents' deserve our keen interest

Minamata disease, which developed into the world's largest pollution-induced health incident, resulted in the death of 1200 people, counting only the officially recognized cases.

The official date of discovery of Minamata disease was 1 May 1956, but cases were cropping up even before that and some people had pointed out from earlier on that abnormalities were seen in cats and other animals that had eaten fish contaminated with mercury.[18] The British neurologist Douglas McAlpine, who saw the true state of the situation in Minamata, published in a medical journal in September 1958 his finding that the characteristic conditions of Minamata disease were entirely similar to the methyl mercury poisoning of workers at a pesticides plant that was reported in the United Kingdom in 1940.[19] Furthermore, a team of researchers studying Minamata disease in the Kumamoto University Department of Medicine also pointed out that the pathological impression was entirely consistent only with the organic mercury poisoning that had been recorded earlier.[20]

Nevertheless, no measures at all were taken by either the chemical industry or the Ministry of Trade and Industry with respect to the liquid industrial wastes from acetylaldehyde manufacturing processes. Because of that, the Showa Denko Kasei factory released waste water containing mercury into the Agano river, contaminating fish and consequently causing the outbreak of Minamata disease in Niigata. It cannot be denied that the outbreak of Minamata disease in Niigata resulted from the failure to regard the occurrence of Minamata disease seriously as an 'incident'.

The same can be said for the disease known as the Itai-Itai disease, which broke out in the Jintsu river basin in Toyama Prefecture. Cases were reported in 1955 and were mainly characterized by kidney damage and osteomalacia (softening of the bones). In 1961 those cases were suspected to have been caused by waste water from the Kamioka mine of the Mitsui Mining and Smelting Corporation.[21]

The Japan Power Reactor and Nuclear Fuel Development Corporation has, by a series of false reports and manoeuvres to conceal the truth, destroyed public trust in the reliability of Japan's nuclear power programme. Even setting this problem aside for the moment, we know at least that the radiation leak accidents, the explosion and fires at the asphalt solidification facility, and so on, had been preceded by the same kinds of situation.[22] Could we not point to the failure to get to the truth of those situations, understanding them as 'incidents', and taking countermeasures as the greatest cause of these accidents?

In the field of economics, too, there are numerous examples of this sort.

In the case where a single trader of the Daiwa Bank lost approximately 1.1 billion dollars, even though improper transactions at the bank's New York branch office had been disclosed, no countermeasures were taken by the senior management.[23] This also can be called a case of ignoring an incident.

In the end, Daiwa Bank had to cease operations in the United States and pay a fine of 340 million dollars. The effect of this incident was not limited to Daiwa Bank as a single enterprise, but rather it spread distrust of Japan's entire financial administration.

Examples like this are not unique to Japan. The prestigious Barings Bank, which has a history of more than 230 years and is said to be the oldest bank in the city of London and is banker to Her Majesty the Queen, is another example. The long history of this prestigious bank was brought to an end by the loss of 900 million pounds in 1995 by one of the bank's traders in Singapore.[24]

The Bank of England investigated ways to rescue Barings, but finally abandoned the idea. Barings was then acquired by the Dutch finance and insurance group, ING. This situation, too, is said to have been preceded by numerous signs of improper transactions.

The early detection of these situations that develop into accidents or scandals and nipping them in the bud are certain paths towards safety. That is to say, what requires our watchful attention is the incidents.

Failure to see these situations, which are buds that will soon grow to a large size, or taking them lightly, to say nothing of concealment or embellishment of information, must be severely admonished.

E.3 IRAS: the key to greater safety

As described in Chapter 1, progress and development in science and technology contribute to human society, but on the other hand necessarily bring about new types of latent danger.

The inertial navigation system (INS), an epoch-making development in aircraft equipment, greatly improved the accuracy of the aircraft attitude indicator and direction display indicator through the use of the gyroscope of the system, lightened the pilot's workload by connection with the autopilot system, and made highly accurate flight navigation possible. On the other hand, however, it has created new kinds of danger, including accidents resulting from the discord between man and machine that accompanies increased automation, such as forgetting to reset the mode switch after manoeuvring to avoid cumulonimbus.

Also, several years after the Boeing 767 was put into service, errors were discovered in the navigation computer program, which controls the autopilot system of the aircraft. Automation and computer control have increased convenience, but this equipment, including its software, has not attained 100 per cent reliability.

Furthermore, it is difficult for the people that depend on automated systems to cope immediately with sudden, unforeseen problems that arise with such systems. It is common knowledge that time and again the air traffic control capability is temporarily paralysed when the radar computer used in air traffic control goes down, leaving the air traffic controllers unable to respond to the situation in an impromptu fashion. No matter how far automation and computer control advance, what ultimately ensures safety will be human judgement and skill.

As we can see from these examples, technological advancement and accident prevention are like the two wheels of a cart; further progress and development of science and technology naturally demand a parallel progress in the establishment of measures for preventing accidents, and it would not be excessive to say that this is a social requirement.

In order to prevent accidents, we must continually search for new types of latent danger factor. Whether the dangers that inevitably come with progress in science and technology can be held to a minimum depends on the minds (ethics) and wisdom of the people who do the science.

Up to now, the statistical yardstick for accident countermeasures has been one occurrence in a million (10^{-6}). However, anticipating new types of danger that cannot be predicted in large and complex systems, for which there can be huge human and economic loss once an accident occurs, it is necessary to strive for an even greater statistical safety margin of one in 10 million or more (10^{-7}).

It must be noted, however, that, if larger and faster aircraft become practical, even if the statistical probability can be lowered, it is not possible to lower the loss of life or the cost of the damage.

IRAS is one method for achieving a higher level of safety, and the soundest method at that.

The important thing with IRAS to begin with is an objective understanding of the true situation of the danger factors. Even simply the feedback of reported incident examples has the effect of making it easier to avert danger when the same or similar circumstances are encountered by accumulating the experiences of other persons as one's own know-how. However, the simple addition of explanations and comments to the collected information with feedback to the relevant persons and organizations does not do justice to the true value of IRAS.

Accordingly, although various methods of analysing human factors have been tried, none of them has been found suitable for the analysis of incident information. What is important in performing IRAS from the viewpoint of preserving safety is the accurate analysis by the MAIR method that was introduced in Chapter 1, and a comprehensive consideration of the results by using quantification method III to obtain an objective understanding of (a) the correlations among the modes of human errors, (b) the correlations between the modes of human errors and the background factors that lead to those errors, and (c) the degrees of danger that are associated with those factors from a macroscopic perspective.

That is to say, the A (analysis) part of IRAS is important. It is then necessary to propose countermeasures for eliminating or avoiding the factors, proceeding in order of their degree of danger. Accordingly, the follow-up investigation of the proposed improvements and measures is also important, because the effectiveness of the proposed measures must be evaluated objectively.

If quantification method III is applied to the results of analysing the incident information before and after improvement measures have been implemented and the results of that application are compared, the distribution of latent danger factors should be different if the measures were effective. It is also necessary to confirm which factors take on high degrees of danger as a result of advances in science and technology.

It is necessary continually to deal with new dangers through repeated investigations, as well as to evaluate the effectiveness of improvement measures. IRAS also holds important significance for that continuity as well.

E.4 Surveys that cannot be done by overseeing authorities and airline companies

Although IRAS is, as described above, a means towards the very important end of preventing accidents, that there are forms that do not work in practice is not simply meaningless; a skeletonized organization is even harmful to safety.

In Japanese aviation, various safety reporting systems have been employed. To begin with, Paragraph 3 of Article 76 of the Civil Aeronautics Law of Japan obligates the captain of the aircraft to report to the Ministry of Transport concerning certain types of incident other than accidents. Examples cited for the content of those reports include 'a malfunction of any of the air navigation facilities or other situations that are deemed to affect the safe operation of aircraft (abnormal weather conditions such as severe turbulence, severe changes in terrestrial and marine phenomena, volcanic eruptions, etc.)'. The incident information that pilots are obligated to report to the Ministry of Transport in this way is limited mainly to 'hard' factors such as facilities, equipment, and weather. There are, however, almost no examples of reports concerning these types of abnormal situation.

Furthermore, Section 2 of Article 76 of the Civil Aeronautics Law of Japan specifies that the captain must report to the Ministry of Transport 'when he has recognized during flight that there was the danger of collision or near miss with another aircraft during flight'. However, the information that must be reported is nothing more than the date, place, and the attitude, altitude, speed, heading, and airspeed of the reporting captain's aircraft.

The near-miss information reported by captains to the Ministry of Transport between 1983 and 1991 amounted to only 12 cases, of which five were confirmed to be near misses, and there were no reports at all for 1992 and 1993. In the 4 years from 1994 to 1997, there were ten reports, of which two were confirmed and four are under investigation.

According to a questionnaire survey conducted by the Flight Crew Unions Federation of Japan in 1986, however, it is clear that at least 72 Japanese pilots of commercial airliners experienced near-miss incidents in 1986 alone. Moreover, a survey of air traffic controllers that was conducted by the Ministry of Transport Labour Union revealed that 288 or more air traffic controllers experienced near-miss situations in the 1 year from June 1986 to May 1987, and that number reached 370 for the year from June 1996 to May 1997.

The difference between these figures and the number of reports to the Ministry of Transport is nothing but astonishing.

The way things are now, the system for reporting abnormal situations and near misses that is mandated by the Civil Aeronautics Law of Japan is not working sufficiently well, as we see from these data. In addition, although the Ministry of Transport has also established a channel for reports of near-miss situations from air traffic controllers, the reality of the matter is that it, too, is hardly functioning.

Each Japanese airline company also has a reporting system referred to as the Captain's Report, which obligates the captain to report any abnormal operation of the aircraft that is encountered during flight.

In the 5 year period from 1983 to 1987, a total of 1405 incidents were reported by means of this system. Of those, most were turn back to the airport after take-off (886 cases) or changes in destination (282 cases). Cases of turn back to the airport after take-off and changes in destination are well understood by external observations, without reporting by the persons directly involved in the cockpit. The fact that these kinds of obvious abnormal operations make up nearly all of the reports obtained through this system can be taken as an indication that the abnormal operations that are known only to the persons operating the aircraft are not included.

Aside from that, Japan Airlines established a company-internal safety reporting system in 1977, and has been collecting information on human-factor incidents.

As that company explains to the reporting persons, 'The valuable information that you submit in this report to contribute to safe operations will not be handled in such a way as to assign responsibility or administer punishment, even if it involves information on your own negligence'. However, no more than a mere 177 reports were filed through this system in the 5 years and 8 months from July 1982 to February 1988. Furthermore, those reports included such matters as persons becoming sick during the flight, and it cannot be said that this system has been successful in collecting information on human-factor incidents that involve pilots and others.

As described in the beginning of this book, accidents that involve nuclear power plants and aircraft, which are large and complex systems, can be called social and public problems because of the scale of human and economic loss that they cause. Now, it is not only insufficient to tackle the issue of safety in large and complex systems from the viewpoint of improving the profits and labour conditions of a single enterprise, it is unsuitable. The reason is that, if the study of safety problems is entrusted to individual companies, there is concern that the self-interests of those companies will be given priority.

Also, in safety reporting systems in the field of aviation, the door must be open to all persons who are involved in the safe operation of aircraft, and, in this sense, IRAS requires a global quality. If this is implemented by the airline companies, however, the problem of reporting information on incidents that involve air traffic controllers, who are under the authority of the Ministry of Land, Infrastructure and Transport, to the airline companies is probably not practical.

E.5 Why is the current system of safety reports not working?

As we see from the preceding discussion, it cannot be said that any of the safety reporting systems that are currently used in Japan are functioning sufficiently well. The clear reason for that failure is that the overseeing authorities and the airline companies are involved in those reporting systems. Setting aside the cases in which there is a relationship of solid trust, is it not simply a matter of human nature for the persons engaged in aviation not to want to take the initiative in reporting their own errors to the Ministry of Land, Infrastructure and Transport, which has the authority of administrative punishment, or to their superiors in the airline company for which they work, which has the authority of personnel management?

It might be mentioned, with respect to this, that Article 30 of the Civil Aeronautics Law of Japan grants the Minister of Land, Infrastructure and Transport the authority to revoke the certification of a person engaged in aviation or to suspend the person for a period of up to 1 year when any one of the following two conditions applies:

1. The person violates this law or a disposition based on this law.
2. The person is guilty of misconduct or gross negligence in doing aviation work.

Incidents involve no major loss, but they include events that occur prior to an accident that may be judged as acts that violate the law. In other words, aviators cannot ignore the possibility that, if they report incidents that involve their own mistakes to a national administrative organization, they may be punished by revocation of their license or suspension from work. It is unthinkable that a person who fears such repercussions would report an incident.

If, for example, drivers of automobiles were asked, for the purpose of contributing to safety on the roads, to report unsafe events to the National Police Agency, which has the authority to revoke or suspend driving licenses, would anyone report their own driving errors even though no penalties would apply?

That this is true is also evident from the unanimous opposition of the Airline Pilots Association (ALPA) and other pilots' organizations to the

Aviation Safety Reporting Programme (ASRP) that the Federal Aviation Agency (FAA) planned to employ in 1975. The pilot opposition stemmed from the fact that the FAA is the overseeing authority of airline companies and has the authority to administer punishment for a violation of the law. In the end, after it was implemented, the programme had to be transferred to the authority of the National Aeronautics and Space Administration.

Accordingly, in the case that national governmental organizations collect incident information, there must be legal measures to ensure the immunity from prosecution and confidentiality of the source of information with respect to the reporters of incident information. However, no matter how valuable the report is to safety, the granting of unconditional immunity for illegal acts is probably not possible because of the need to maintain parity with other fields (such as nuclear power generation or medicine). Even assuming for the moment that it were possible to implement such a system, we would have to wonder whether or not the persons reporting incidents would trust the government organizations.

Also, even assuming that an internal safety reporting system that is implemented by an airline company can smoothly provide a large amount of information, would that be effective for the purpose of preserving safety in the sphere of aviation as a whole? If the data obtained are circulated only within the company that collected them and not passed on to other companies that fly the same type of aircraft or the same air route, then it is not too much to say that the value of that information will be reduced by half.

A vivid demonstration of this is an accident involving TWA (Trans-World Airlines) flight 514 Boeing 727 that occurred on 1 December 1974. When TWA flight 514 was making a VOR (VHF Omnidirectional radio range) approach for landing at Dulles International Airport in Washington, DC, its altitude was too low and it crashed into a hill in front of the airport, killing all 92 of the passengers and crew on board.[25]

In a VOR approach, the aircraft descends to the minimum descent altitude in the final approach course while receiving a radio signal that is transmitted horizontally by a radio beacon within the airport. While maintaining that minimum altitude, the pilot makes a visual sighting of the runway and then lands the plane. Flight 514, however, mistook the minimum altitude and was flying too low, and crashed into the hill as a result. As the cause of the accident, a misunderstanding of a term used in the communication with the air traffic controller has been pointed to.

However, the same kind of situation as had occurred in this disaster had also been experienced just 6 weeks earlier in an incident involving a United Airlines plane. While the pilot of the United flight was making a VOR approach for landing at Washington Dulles Airport, there was a mutual misunderstanding in the communication with the air traffic controller that resulted in the plane descending to an altitude below the minimum altitude before reaching the hill in front of the airport, and the plane nearly collided with the hill.

The United Airlines pilot, believing that a single misstep would have resulted in a disaster in this situation, reported the incident using the company's safety reporting system. Upon receiving the report, United Airlines issued a 'cautionary letter of advice' to all of its operations crew members to call their attention to the incident. Just after that, the TWA accident occurred, having exactly the same cause as was reported within the United Airlines reporting system.

Although that accident is said to have been the opportunity for the Federal Aviation Agency to introduce the Aviation Safety Reporting Programme (ASRP), it is well known that the scope of the usefulness of valuable incident information is narrowly restricted in the case of safety reporting systems that are limited to within ordinary individual airline companies.

E.6 Requirements for establishing IRAS

From the above discussion, neither government agencies nor airline companies are suitable for the implementation of IRAS.

So then, if we ask what kind of organization should take up that implementation, it can be none other than a private third-party organization whose mission is only the prevention of accidents and the preservation of safety. The nature of this private third-party organization must be neutral and scientific.

Concerning the human factor incident reports that are the key to the success or failure of IRAS, naturally the reports would not be made without a high awareness of safety on the part of the reporter. At the same time, however, the truth of the situations would not be brought to light unless the reports are made voluntarily and on the basis of a strong relationship of trust with the organization that receives the reports.

Then, in order to secure that relationship of trust, it is not sufficient for that organization simply to be formally independent of the organization to which the information reporters belong; it must of course be essentially and completely independent on the personal and economic levels. It stands to reason that there should be no participation by persons who might have an influence on the employment of the reporting individual, or on punishments or other consequences to the reporting individual.

Frankly, the companies concerned or their affiliated organizations, organizations that are auxiliary to government agencies and organizations in which high-ranking officials of overseeing governmental agencies or senior executives of the companies concerned participate, including retired executives, are not appropriate organizations for implementing IRAS.

In the sphere of aviation in Japan, there are public service corporations for which the Ministry of Land, Infrastructure and Transport serves as the relevant authority. If IRAS is implemented by a new or existing such corporation, it would not be possible to maintain essential autonomy and neutrality, because the work of that corporation would be overseen by the Ministry of Land, Infrastructure and Transport, and, as the competent authority, the Ministry could issue orders required for oversight and at the same time have the authority to inspect the operations and assets of the corporation (Article 67 of the Civil Code). It is thus clear that the relationship of trust that is the foundation for IRAS cannot be established in that case.

It might be mentioned here in passing that in September 1987 the Flight Crew Unions Federation of Japan, to which nearly all of the pilots of Japanese airline companies belong, had already passed a resolution to 'oppose safety reporting systems implemented by the Ministry of Land, Infrastructure and Transport or other such organizations and to work against the establishment of such systems'. This matter was brought up by a member in the 112th Meeting of the House of Councillors on the Audit Committee (25 April 1988).[26]

In a recent survey conducted by the Japan Federation of Air Safety and published on 12 April, the 30 500 members of 58 labour unions, including the Ministry of Transport Union and the Meteorological Workers Union and others, were asked their opinions concerning the safety reporting system that is being set up by the Ministry of Transport. The responses were 20 per cent in favour, 67 per cent opposed, and 12 per cent responded with 'other', thus 80 per cent were not in favour. Nevertheless, the Ministry of Transport requested that the Japan Civil Aviation Promotion Foundation investigate the introduction of the safety reporting system.

The main reason for the opposition is the belief that the organization that is to operate that system is not a completely independent third party because it is a government organization that has jurisdiction over aviation.

In addition to that, there is the belief that the organization that is to operate the system is linked to government administration organizations that have the authority to punish.

With this description, member of the House of Councillors Hisamitsu Sugano inquired of the administrative committee member (the head of the Ministry of Transport's Civil Aviation Department) what he thought of the matter.

To that, the committee member responded as follows:

The fundamental issue with respect to this system is that it first of all function well, so, from that point of view, I suppose the results of that questionnaire are possibly somewhat meaningful.

Accordingly, we of course do not consider that the Ministry of Transport, as a government organization, would directly administer the system, and, although the airline companies currently have internal reporting systems, those do not work sufficiently well. It is thus also probably not suitable for an organization within the airline companies or one over which they have authority to administer the system. We therefore believe that some third party that is independent from those organizations must administer the system.

A person affiliated with JAPAN ALPA, which is a member of the international organization of civil airline pilots, IFALPA (International Federation of Air Line Pilots' Associations), answered that, even now, the understanding of the persons making reports does not differ greatly from the above.

If the system is not administered by an independent organization, it will not be possible to distribute information to the individual airline companies laterally and rapidly and it will not be possible to prevent tragedies such as the crash of the TWA (Transworld Airlines) flight 512 described earlier.

After that, in May 1996, the US Federal Aviation Agency formulated a plan for the Global Analysis Information Network (GAIN), an international organization whose purpose was the collection and exchange of information for aviation safety, and, in May 1997, the Conference Preparation Committee (CPC) was set up for examination.

The members of that same committee were officials and employees of the US Federal Aviation Agency, NASA, the International Air Transport Association (IATA, an international organization of commercial airline companies), the Regional Airlines Association (RAA), the Flight Safety Foundation (FSF), the US Navy, the Boeing

Corporation, the Airbus Corporation, ALPA, and other such organizations. This committee will continue to study the organization and specific functions of GAIN.

Also, there have recently been various attempts to use e-mail and the Internet to collect incident information that is related to aviation safety.

As described earlier, however, the safety reporting system devised by the US Federal Aviation Agency, the government organization that has authority over the persons who make the reports, had to be turned over to NASA for administration after it was inaugurated in 1975. Also, in July 1985, IFALPA set up the IFALPA ASRS, premised on integration with the NASA safety reporting system (the Aviation Safety Reporting System, ASRS) and called on pilots to provide incident reports. However, even several years later, not a single report had been submitted from any of the national associations in the federation.

Considering that, the noteworthy question is whether or not GAIN can obtain the trust of the persons who report information so that it is possible to collect reports that contain highly accurate and true-to-the-matter incident information involving the reporters' own errors that has not been brought to the surface. Collection by the use of e-mail and the Internet, too, regardless of the means of collection, whether or not there is fear that the reporter may suffer some disadvantage, and whether or not the information that is reported is made good use of, is a key to future development.

No matter how excellent the organization appears to be or how convenient the means of reporting, if there is no basic relation of trust, then we cannot expect the human-factor incident reports to describe the truth concerning human factors.

Safety is something that essentially requires strict impartiality, so naturally it should not be affected by ideology or other considerations (national advantage, corporate profits, or benefit to a certain point of view). Because a high degree of specialized knowledge is required in the field of aviation, in the past this has been a field in which abuse of authority on the part of the experts as a group could easily occur, with decisions being made arbitrarily in such a way as to prevent laymen from approaching, and it has been very difficult to achieve an effect of checks and balances. Those involved in the sphere of aviation should probably humbly reflect on this point. Also, as mentioned repeatedly in this book, the actual facts of the situation must, to the utmost, be confronted with a humble attitude.

Therefore, it is essential that IRAS be administered by a non-governmental third-party organization that has a neutral and scientific nature, regardless of the field. Nevertheless, although the civil aviation industry could obtain the most benefit from an effectively functioning IRAS – that is, the implementation of IRAS by an absolutely neutral third party organization – the industry generally does not have active support for this situation. It may be that a latent hesitation on the part of a person who has made an error to leak information concerning incidents that are internal to the person's company or industry to an outside organization (i.e. a third-party organization) is at work.

Also, as described earlier, not only is the administration of IRAS by the government administrative organ that has oversight authority over the person who is making the report unreasonable to begin with, but, looking at the field of aviation, the Minister of Land, Infrastructure and Transport, which establishes and administers airports, is a central figure in establishing air traffic control facilities and personnel, and the situation is the same as for the airline companies taking a latent wrongdoer position in many air accidents. Government does not generally take a highly positive stance in relation to the implementation of IRAS by a private third-party organization that takes on an ombudsman-like existence.

There is, here, a major problem in the strategy against accidents.

As a condition for the implementation of an IRAS that truly works, what is required first of all is a major reform in the general social awareness and understanding of the safety issue, beginning with the government and the various industries.

IRAS is a crucial key to the prevention of accidents.

Appendix A

Explanation of quantification method III

The development of this theory will firstly be explained, because it is useful in the application of this method.

We will illustrate the results of a survey using Tables 9 and 10, which depict how respondents (denoted by is) select responses. The response categories are shown for all questions using symbols such as L_1, L_2, ... L_R, with each serial number denoting the consecutive response categories to the items in question. The check marks in the tables indicate the response categories selected by each respondent for each question.

If we assume that people with similar check mark patterns are grouped together and that categories that are selected by similar people are grouped together – in other words, if we assume that it is possible to classify respondents and categories simultaneously – then it becomes possible to clarify the manner in which respondents are close together or distant with respect to how they respond. We can then change the arrangement of respondents and response categories so that related categories are grouped together and so that respondents who show similar response patterns are also grouped together. This attempt was intended to produce a clear locational correspondence between respondents and categories. If we can produce such a table, then we will have accomplished our purpose.

To be more concrete, let L be a good thing. A good thing (check mark) is selected by a person of refined taste, and a person of refined taste is the one who chose the good thing of good taste. On the surface, this process seems like synonymous repetition, but we can achieve our purpose if we classify respondents and good things simultaneously, thus grouping similar respondents and similar good things and assigning names to them. Let us think about this by referring to Tables 11 and 12.

The primes in Table 11 indicate the responding entries from Tables 9 and 10 in the manner discussed above. By rearranging the table in this way, those categories that are mutually selected at the same time are located close to one another, while those categories that do not have mutual relationships and that do not have check marks (i. e. those not mutually selected, at the same time) are located distant from one another. Looking at this from the respondents'

Table 9 Respondent response patterns

Subject	Object Response category						
	L_1	L_2	L_3	...	L_{Jj}	...	L_R
Respondent							
1	V		V				
2		V	V				V
3	V		V				
4	V				V		
:							
Jj	V	V	V		V		
:							
Q		V			V		V

A check mark indicates that a certain respondent has the characteristics L_i. Q is the total number of patterns that a respondent shows. If all respondents show different patterns, then the frequency is identical to the total number of respondents.

Table 10 Specific example for R (number of categories) = 6 and Q (number of respondents) = 8

Subject	Object Response category					
	L_1	L_2	L_3	L_4	L_5	L_6
Respondent						
1	V			V		V
2	V		V	V		
3				V		V
4	V		V		V	
5		V				V
6		V		V		V
7	V					V
8	V		V			

Object = label; subject = individual.

side, those types of respondent who are located close to one another show similar reaction patterns, while those types of respondent who are distant from one another show different response patterns. Therefore, through this simultaneous rearrangement, it becomes possible to unite similar things (in this case, respondents and categories).

The related categories are drawn close together and the unrelated categories come to be located at both poles of the first dimension. In this way, we can obtain configurations of categories that move gradually from 'close' to 'distant'.

If the similarity of the above-mentioned response patterns is due to a

Table 11 Response patterns

Subject	Object Response category						
	$L_{1'}$	$L_{2'}$	$L_{3'}$	$L_{4'}$	$L_{5'}$	...	$L_{R'}$
Respondent							
1'	V	V	V	V	V		
2'		V	V			V	
3'			V		V	V	
4'				V			
⋮							
Q'						V	V

Table 12 Concrete example of Table 10, $R=6$, $Q=8$

Subject	Object Response category					
	L_5	L_3	L_1	L_4	L_6	L_2
Respondent						
4	V	V	V			
2		V	V	V		
8	V	V				
1			V	V	V	
7			V		V	
3				V	V	
6				V	V	V
5					V	V

connection that can be expressed efficiently in the first dimension, we can summarize the information on similarities among individuals and response categories on one axis. However, even in this case (expressing the first dimension), the rearrangement becomes quite difficult when the number of types, Q, and the total number of categories, R, are large.

Another approach to this arrangement is facilitated by the analytic method. This becomes possible using the concept of the correlation coefficient.

Relation between quantification method III and correspondence analysis

This method has a curious history. It has been many years since the method was developed, but it is not so popular in English-speaking countries, although it is very useful for exploratory data analysis, while it is very popular and frequently used in various fields in Japan and Europe. I suppose that this is mainly due to the fact that this method has been developed on the basis of the original 'anti-test-estimation-mathematical statistics' philosophy of data analysis in Japan or France, apart from the mainstream of so-called mathematical statistics where only statistical tests and estimations have been referred to as 'statistical methods' in English-speaking countries. Actually, this method is not discussed in mathematical statistics but in psychometrics, with the data analytic method of classification or clustering, in the United States. This situation will be improved by an increase in English books describing this method.

Here, I should like to touch briefly on the relation between quantification method III or quantification of response pattern and correspondence analysis. This method of quantification was published in 1956 in Japan under the term 'quantification of response pattern', which would later be called 'quantification method III' by a user, Professor H. Akuto in social psychology, and has been applied in various fields since then.

On the other hand, correspondence analysis was independently published by Professor J. P. Benzecri, in 1973 in France, and has been used in various fields. The origins of these methods are remarkably similar and the processes of application are also quite similar. The contents of these methods are formally equivalent, although the leading idea may be somewhat different. Such being the case, this method is called 'quantification method III' or 'quantification of response pattern' in Japan, while it is called 'correspondence analysis' in Europe. In a word, this method represents the dissimilarity of response categories in such questions as a distance in Euclidean space based on many cross-tabulations between two questions. A larger distance corresponds to a high degree of dissimilarity realized by graphic representation in Euclidean space. [See Hayashi, C. (1992) Quantification method III or correspondence analysis in medical science. *Annals of Cancer Research and Therapy*, **1**(1).]

There follows an outline of the computation formulae used in quantification method III, as shown in quantification manipulation form

$$^1\rho = \frac{C_{xy}}{\sigma_x \sigma_y}$$

Now, before we determine the maximum value of this correlation coeffi-cient $^1\rho$, to obtain this from the data we must first define our symbols:

$\delta_i(j) = 1$ if the ith individual (type) selects the jth response category (having a check mark)

$= 0$ otherwise (no check mark)

$\delta_i(j)$ $(i = 1, 2, ..., Q; j = 1, 2, ..., R)$ can be obtained from the survey data. Let l_i

be the number of items that i selects, then

$$l_i = \sum_{j=1}^{R} \delta_i(j)$$

Parameter s_i is the number of persons who show check mark patterns the same as the ith type (shown in Table 13). In other words, s_i is the number of respondents who fall into the ith type. If we let n be the sample size, then

$$n = \sum_{i=1}^{Q} s_i$$

and Q be the total number of types shown by check marks, then, using these signs, we define σ_x^2, σ_y^2, and C_{xy} by including the values of the data

$$\sigma_x^2 = \frac{\sum_{i=1}^{Q} \sum_{j=1}^{R} \delta_i(j) s_i x_j^2}{\ln - \left\{ \frac{\sum_{i=1}^{Q} \sum_{j=1}^{R} \delta_i(j) s_i x_j}{\ln} \right\}^2}$$

$$\sigma_y^2 = \frac{\sum_{i=1}^{Q} s_i l_i y_i^2}{\ln - \left\{ \frac{\sum_{i=1}^{Q} s_i l_i y_i}{\ln} \right\}^2}$$

$$C_{xy} = \frac{\sum_{i=1}^{Q} \sum_{j=1}^{R} \delta_i(j) s_i x_j y_i}{\ln}$$

$$- \left\{ \frac{\sum_{i=1}^{Q} \sum_{j=1}^{R} \delta_i(j) s_i x_j}{\ln} \right\} \left\{ \frac{\sum_{i=1}^{Q} s_i l_i y_i}{\ln} \right\}$$

To maximize $^1\rho$, the next equations to be solved are

$$\frac{\partial^1 \rho}{\partial x_k} = 0, \qquad \frac{\partial^1 \rho}{\partial y_e} = 0 \qquad (k = 1, 2, ..., R; \ e = 1, 2, ..., Q)$$

From these we obtain

$$\sum_{j=1}^{R} h_{jk} x_j = {}^1\rho^2 \left(d_k x_k - \sum_{j=1}^{R} b_{jk} x_j \right)$$

Table 13 Response patterns of individuals to questions

			A	1			2	...		L	
			B	1	2	...	K_1	K_1+1	...	...	...	R
			C	C_{11}	C_{12}	...	C_{1k1}	C_{21}	...	C_{2k2}	...	C_{L1}	...	C_{LKi}
D	E	F									
I_1	s_1	1	✓				✓				✓
I_2	s_2	2	✓				✓				✓
I_3	s_3	3		✓					✓	✓	
.	.	.									
.	.	.									
I_Q	s_Q	Q	✓						✓		✓

A = question; B = consecutive number; C = response category; D = total of signs; E = frequency; F = response type (individuals).

Note: The check mark denotes the response category selected by the individual. In general, the response categories do not include 'don't know' or 'other' response options. Because of this, respondents do not always necessarily react to the categories given for each question. As a consequence, the number of responses, 1, differs among questions. If we include 'other' and 'don't know', 1 always equals L (the number of questions). In this case, where individual reactions are being delineated, s is always 1 and Q becomes the total number of samples. It is not accurate to express types because there are some who have the same reactions. This table summarizes the individual reactions in their original form.

where

$$h_{jk} = a_{jk} - b_{jk}$$

$$a_{jk} = \sum_{i=1}^{Q} \frac{\delta_i(j)\delta_i(k)}{l_i} s_i$$

$$b_{jk} = \frac{1}{\ln} \sum_{j=1}^{Q} \delta_i(j)s_i \cdot \sum_{i=1}^{Q} \delta_i(k)s_i$$

$$d_k = \sum_{i=1}^{Q} s_i\delta_i(k), \qquad \ln = \sum_{i=1}^{Q} l_i s_i$$

Furthermore, by modifying this equation using f_{jk}, we obtain

$$\sum_{j=1}^{R} h_{jk}x_j = {}^1\rho^2 \sum_{j=1}^{R} f_{jk}x_j \qquad (k = 1, 2, ..., R)$$

where

$$f_{jk} = -b_{jk} \qquad \text{if } j \neq k$$

$$f_{jk} = d_k - b_{jk} \qquad \text{if } j = k$$

For convenience in later discussion, the matrix representation $\mathbf{HX} = {}^1\rho^2\,\mathbf{FX}$ will be used, where the elements of matrix $\mathbf{H}$ are h_{jk}, those of matrix $\mathbf{F}$ are f_{jk}, and X is a column vector. By solving this equation (i.e. $\mathbf{HX} = {}^1\rho^2\,\mathbf{FX}$), we obtain the maximum value of ${}^1\rho^2$, and the value of x that corresponds to this. Of course, as a mean can be determined arbitrarily, we solve the equation by arbitrarily making $x_i = 0$ (for example, $x_R = 0$). Also, $\mathbf{H}$ and $\mathbf{F}$ are symmetrical. Instead of letting $X_R = 0$, we can set the total means to 0, so that $\bar{x}$ can be set to 0, shown as

$$\bar{x} = \frac{1}{\ln}\sum_{i=1}^{Q}\sum_{j=1}^{R}\delta_i(j)s_ix_j = 0$$

Then we can obtain

$$\sum_{j=1}^{R}a_{jk}x_j = {}^1\rho^2 d_kx_k \qquad (k = 1, 2, ..., R)$$

We solve this equation by assuming that

$$\bar{x} = 0$$

If we set

$${}^1\rho^2 \neq 1$$

then

$$\ln\bar{x} = \sum_{k=1}^{R}d_kx_k = 0$$

Therefore, to satisfy this equation is always to satisfy the condition $\bar{x} = 0$. It is convenient to perform numerical calculations by using this approach. In other words, this is equivalent to solving the equation

$$\sum_{j=1}^{R}h_{jk}x_j = {}^1\rho^2 d_kx_k$$

as it is. It is found that the solution always satisfies the above-mentioned equation. On the other hand, as the value of y is

$$y_e = \frac{1}{{}^1\rho}\frac{\sigma_y}{\sigma_x}\left(\frac{1}{l_e}\sum_{j=1}^{R}x_j\delta_e(j)\right)$$

it is easily obtained from the value of x. As size can be set arbitrarily, we can assume that

$$\frac{1}{l_\rho} \frac{\sigma_y}{\sigma_x} = 1$$

From this

$$y_e = \frac{1}{l_e} \sum_{j=1}^{R} x_j \delta_e(j)$$

Here, we had best take the mean value of the selected x_j. In the present case we assign y for 'respondent' and x for 'category'. This is merely one way of doing this. We could assign x to 'respondent' and y to 'category' as they are both the same. The important thing is to make the dimension $(R-1)$ as small as possible when solving the above-mentioned equation. The thing to be cautious about, in using the above-mentioned equation, is l_i. If we include 'other,' 'don't know', and 'no response' categories for a certain question, then l_i will always equal the number of questions. However, the exclusion of 'other,' 'don't know', and 'no response' from the response categories of a certain question is sometimes desirable for clear understanding of the phenomenon. If we eliminate these from the categories to be taken up for computation (i.e. excluding them from $L_i,...,L_R$), then l_i is different from one respondent to another. This is the same as in the case where there are missing data. If we include 'other' and 'don't know' in the calculation, the axis that distinguishes those who responded from those who did not will always emerge.

Now, what should we do in a case where we cannot rearrange the data well in one-dimensional space? In this case we attempt to rearrange the respondents or categories in two-dimensional space. It is best to rearrange them as shown in Fig. 24 which clusters categories and respondents. If we cannot rearrange them well in two-dimensional space, then we can do a better job in three-dimensional space. We use the analytic method for this approach.

In the case of two-dimensional space, let us assign y_i, v_i $(i=1,...,Q)$ for an individual and let us assign x_j, u_j $(j=1,...,P)$ for a category. We want to quantify individuals (types) or categories by assigning numerical vectors to them to minimize them within generalized variance, $| \mathbf{W} |$, with the total variance being constant. In other words, the intent is to minimize $| \mathbf{W} |/| \mathbf{VT} |$, where $| \mathbf{VT} |$ is the generalized total variance with respect to vector x_i (or y_i) for $s = 1,2,...,s$ and for all i (or j), with S being the number of spatial dimensions. This process is described in detail below.

We consider maximizing $1 - | \mathbf{W} |/| \mathbf{VT} |$ under the reasonable condition that the non-diagonal elements in matrix $\mathbf{W}$ vanish; this means maximizing

$$\frac{1 - |\widetilde{\mathbf{W}}|}{|\mathbf{VT}|}$$

Two-Dimensional Correspondence

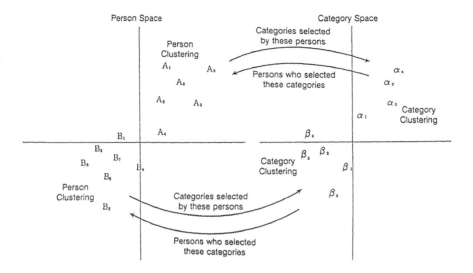

Fig. 24 Mapping diagram: quantification of response pattern classification method

where $\widetilde{\mathbf{W}}$ is the diagonal matrix of $\mathbf{W}$. As

$$\frac{|\widetilde{\mathbf{W}}|}{|\mathbf{VT}|} \geq \frac{|\widetilde{\mathbf{W}}|}{|\widetilde{\mathbf{VT}}|} \qquad \text{and} \qquad |\mathbf{VT}| \leq |\widetilde{\mathbf{VT}}|$$

hold, it follows that

$$\frac{1 - |\widetilde{\mathbf{W}}|}{|\widetilde{\mathbf{VT}}|} \geq \frac{1 - |\widetilde{\mathbf{W}}|}{|\mathbf{VT}|}$$

where $\widetilde{\mathbf{VT}}$ is the diagonal matrix of $\mathbf{VT}$. Thus, it is desirable to quantify the respondents and categories [in other words, to require vector x_i (or y_i) for all i (or j)] so as to minimize

$$\frac{|\widetilde{\mathbf{W}}|}{|\widetilde{\mathbf{VT}}|}$$

or to maximize

$$\frac{1 - |\widetilde{\mathbf{W}}|}{|\widetilde{\mathbf{VT}}|}$$

This reduces to the maximizing of $\coprod_s^S {}^s\eta^2$ for ${}^s\eta^2$ is the correlation ratio with respect to $^s x_i$ for all i, which is equivalent to maximizing

$$\prod_s^S {}^s\rho^2$$

where $^s\rho$ is the correlation coefficient between $^s x_i$ and $^s x_j$ for all i and j. (In other words, this is equivalent to seeking to obtain vectors X, Y, U, and V so as to maximize the area of the rectangle created by $^1\rho$ and $^2\rho$.) This leads us to solve the latent equation $\mathbf{H}X = \rho^2\mathbf{F}X$ to obtain vector ^sX which corresponds to the sth largest latent root of $\mathbf{H}$ which is equivalent to X. By using this procedure we obtain the second largest latent root of $\mathbf{H}$ which is equivalent to U. It is obvious that, between vectors ^iX and ^jX, which satisfy $\mathbf{H}X = \rho^2\mathbf{F}X$, there is the relation $(\mathbf{F}^iX, {}^jX) = 0$. It is found that the solution to this equation satisfies all our conditions, and it is easily seen that the diagonal elements of the within-variance that is obtained as a solution become 0.

When we deal with cases involving more than two-dimensional space, we repeat the above process using the same logic and methodology. In these cases it is better to obtain vectors that maximize the value of $\prod_s^S {}^s\eta^2$ which is the product of the correlation ratios of S. This can be accomplished by determining the largest root for ρ^2, then the second largest root, and so on, by solving the equation $\mathbf{H}X = \rho^2\mathbf{F}X$ to obtain the relevant vector. [See Hayashi, C., Suzuki, T., and Sazaki, M. (1992) *Data Analysis for Comparative Social Research: International Perspectives*].

The computations required for this process can be executed easily using Program Package for Social Science II.

Appendix B

List of accidents

January 10, 1954. Over the Mediterranean off Elba. Comet (BOAC flight 781) that had taken off from Rome disintegrated during climb over Elba Island in the Mediterranean Sea. See page 10.

April 8, 1954. Near Stromboli, Italy. Comet (South African Airways flight 201) aircraft disintegrated during flight and crashed near the island of Stromboli, Italy. See page 10.

September 29, 1957. Chelyabinsk, former Soviet Union. Explosion at a radioactive waste storage facility. See page 4.

May 23, 1960. Boston, Massachusetts, United States. Delta Airlines CV-880 trainer crashed when taking off from a Boston airport for a check flight. See page 106.

October 4, 1960. Near Boston Airport, Massachusetts, United States. Eastern Airlines Lockheed Electra 188A four-engine propeller aircraft crashed. See page 187.

February 27, 1965. Iki Airport, Japan. Japan Airlines CV-880 trainer crashed and burned just short of the runway. See page 106.

August 1966. Tokyo International Airport, Japan. Japan Airlines CV-880 trainer ran off a runway during a take-off run and caught fire. See page 106.

June 24, 1969. Moses Lake Airport, Washington, United States. Japan Airlines CV-880 trainer crashed after take-off during training at the airport. See page 106.

July 27, 1970. Naha Airport, Japan. Flying Tigers air freight flight DC-8 crashed into the sea 2200 ft ahead of the runway. See page 84.

July 30, 1971. Shizukuishi, Japan. Midair collision between the All-Nippon Airways flight 58 aircraft B-727–281 and the Japanese Self-Defence Forces training fighter North American F-86F. See page 7.

October 2, 1971. Belgium. British-European Airlines (BEA) Vickers Vanguard aircraft flight 706 crashed owing to a rapid decompression. See page 12.

December 20, 1972. Chicago O'Hare International Airport, Illinois, United States. Delta Airlines flight 954 CV-880 and North Central Airlines flight

575 DC-9 collided on the ground while moving on the same runway under low visibility in a dense fog. See page 28.

December 29, 1972. Miami Airport, Florida, United States. Eastern Airlines 401 L-1011 crashed when the crew unintentionally deactivated the automatic control system because of excessive focus on a mechanical failure. See page 104.

September 11, 1974. Douglas City Airport in Charlotte, North Carolina, United States. Eastern Airlines flight 11 DC-9 crashed into a field on the approach course and burst into flames. See page 29.

December 1, 1974. Dulles International Airport, Washington, DC, United States. TWA (Trans-World Airlines) flight 514 B-727 crashed into a hill in front of the airport. See page 231.

June 24, 1975. Kennedy International Airport, New York, United States. Eastern Airlines flight 66 B-727 encountered wind shear, struck approach lights, and crashed during landing. See page 92.

November 12, 1975. Raleigh/Durham Airport, North Carolina, United States. Eastern Airlines flight 576 B-727 slammed into the ground during an ILS approach to the airport. See page 101.

December 16, 1975. Anchorage International Airport, Alaska, United States. Japan Airlines flight 422 B-747 skidded on an icy taxiway on the way to the take-off position owing to a strong crosswind and slid down a slope. See pages 27 and 87.

March 27, 1977. Tenerife Airport in the Canary Islands, Spain. KLM (Dutch Airlines) flight 4805 B-747 and PAA (Pan American Airlines) flight 1736 B-747 collided on the runway of Tenerife Airport. See page 18.

May 14, 1977. Near Lusaka, Zambia. IAS cargo B-707 crashed just before landing. See page 198.

April 20, 1978. Murmansk, former Soviet Union. Korean Airlines flight 902 B-707 deviated off course and was forced to land on a frozen lake near Murmansk after being fired upon by a Russian fighter. See page 70.

June 2, 1978. Osaka Airport, Japan. Japan Airlines flight 115 B-747 – the tail of the aircraft struck the runway. See page 9.

December 31, 1978. Beloyarsk, former Soviet Union. Accident involving the No. 3 reactor of the Beloyarsk nuclear power plant. See page 4.

February 12, 1979. Near Clarksburg, West Virginia, United States. Allegheny Airlines flight 561 Nord 262 crashed after take-off from Benedum Airport. See page 88.

March 28, 1979. Three Mile Island, Pennsylvania, United States. Radiation leak at the Three Mile Island nuclear power station. See pages 9 and 38.

May 25, 1979. Chicago, Illinois, United States. American Airlines flight 191 DC-10-10 crashed after an engine fell off. See page 38.

January 13, 1982. Potomac river, Washington, DC, United States. Air Florida flight 90 B-737-222 crashed into the railing of a bridge over the Potomac river. See page 88.

February 9, 1982. Near Haneda Airport, Japan. Japan Airlines flight 350 DC 8-61 crashed into Tokyo Bay owing to the action of a mentally-disturbed pilot. See pages 7 and 36.

September 1, 1983. Sea off Sakhalin Island, former Soviet Union. Korean Airlines flight 007 B-747 was shot down by a Soviet Sukhoi-15 in the sea off Sakhalin Island. See page 70.

December 7, 1983. Barajas Airport, Madrid, Spain. Avianca Airlines domestic flight 134 DC-9 collided on the runway with the Iberia Airlines flight 350 B-727. See page 97.

April 19, 1984. Naha Airport, Japan. Japan Asia Airways flight EG292 DC-8 encountered wind shear and hit a landing approach light. See pages 40 and 93.

December 2, 1984. Bhopal, India. Explosion at the Union Carbide India pesticide plant. See page 2.

May 28, 1985. Naha Airport, Japan. Japanese Air Self-Defence Forces MU-2 and All-Nippon Airways flight 81 B-747 intersected and grazed. See page 30.

June 23, 1985. Atlantic Ocean off Shannon, Ireland. Air India flight 182 B-747 crashed into the ocean (owing to rapid decompression by an explosion). See pages 11 and 15.

August 12, 1985. Near Osutaka Mountain, Japan. Japan Airlines flight 123 Boeing 747SR crashed into a steep mountain ridge. See page 5.

November 30, 1985. Over Sea of Japan. The pilot of Japan Airlines flight 441 B-747 forgot to return the mode selection switch to INS mode after flying to avoid cumulonimbus clouds and the aircraft flew off course, causing the scrambling of Soviet military aircraft. See page 41.

April 26, 1986. Chernobyl, former Soviet Union. Nuclear power plant accident. See page 3.

May 12, 1986. Kai Tak Airport, Hong Kong. Japan Airlines flight 65 B-747 encountered heavy rain during landing approach and the plane landed 500 ft short of the intended touchdown point. See page 84.

November 28, 1987. Mauritius. South African Airways flight 295 B-747-combi crashed into the sea about 150 miles north-east of Mauritius. See page 11.

January 10, 1988. Yonago, Japan. Toa-kokunai Airlines flight 670 YS-11 overran the runway and plunged into Lake Nakanoumi during take-off in light snow. See page 40.

April 28, 1988. Near Honolulu, United States. Part of the ceiling of Aloha Airlines flight 243 B-737-297 blew off in mid-flight. See page 178.

July 23, 1988. Near Miura Peninsula, Japan. Japan Self-Defence Forces submarine Nadashio collided with a fishing boat when the submarine was manoeuvring to avoid a yacht. See page 172.

January 9, 1989. Helsinki Airport. Finland Airlines Airbus 300 B4-203FF – serious incident during approach. See page 42.

February 24, 1989. Honolulu, United States. Accident involving United Airlines flight 811 B-747 out of Honolulu, in which the door fell off during the flight. See page 11.

January 25, 1990. Cove Neck, New York, United States. Avianca Airlines (Columbia) flight 52 B-707-321B crashed into a hill 15 miles north of the airport. See page 102.

December 3, 1990. Detroit Airport, Michigan, United States. Northwest Airlines flight 1482 DC-9-14 collided with Northwest Airlines flight 299 B-727 in dense fog near a runway intersection. See page 97.

February 11, 1991. Moscow Airport. Airbus 310 – serious incident during approach. See page 42.

January 20, 1992. Strasbourg Airport, France. French Air Inter flight 5148 Airbus 320-111 aircraft crashed and burned on approach to Strasbourg Airport. See page 81.

April 26, 1994. Nagoya Airport, Japan. China Airlines flight 140 Airbus 300-600 crashed during landing. See page 41.

June 29, 1994. Over Caribbean. American Airlines flight 107 MD-11 experienced severe up-and-down movement. See page 45.

September 24, 1994. Orly Airport, France. Airbus 300 – serious incident during approach. See page 42.

December 8, 1995. Fukui Prefecture, Japan. Leak at the Monju fast breeder reactor. See page 3.

April 14, 1997. Fukui Prefecture, Japan. Radiation leak at the Fugen reactor. See page 4.

June 8, 1997. Over Mie, Japan. Japan Airlines flight 706 MD-11 experienced severe up-and-down movement during the descent over Mie. See page 44.

January 31, 2001. Over Yaizu, Japan. Near-miss incident at a flight level of approximately 35 000 ft involving Japan Airlines flight 907 B-747-400 and Japan Airlines flight 958 DC-10. See page xv.

July 1, 2002. Over Lake Constance, Germany. Bashkirian Airlines charter flight BTC2937 Tupolev 154 collided with DHL flight DHX611 B-757 (cargo plane) at a flight level of 35 000 ft. See page x.

Bibliography

This book was written not only for specialists around the world but also for a wide range of general readers with an interest in preventing accidents before they occur. It is based on the following books.

Miyagi, M., Incident Reporting System ni tsuite no Shikouteki Kenkyu [Preliminary research on the incident reporting system]. *Review of Air Law and its Practice*, 16–17, pp.1–184 (in Japanese). Tokyo: Japan Research Institute of Air Law (sold by Yuhikaku Publishing Co., Ltd), February 1986.

Miyagi, M., *Koukuu ni Okeru Incident Reporting System ni Kansuru Sougouteki Kenkyuu* [Integrated Study for the Aviation Incident Reporting System – Analysis of Incident Experienced by Flight Crew – Aircraft Operation] (in Japanese). Tokyo: Japan Research Institute of Air Law (sold by Yuhikaku Publishing Co., Ltd), June 1988.

Miyagi, M., *Fukuzatsu Daikibo System ni Okeru Jiko Boushi (I) – Koukuu Koutsuu Kansei Gyoumi wo Megutte* [Accident Prevention in Large and Complex Systems (I) – Air Traffic Control] (in Japanese). Tokyo: Japan Research Institute of Air Law (sold by Yuhikaku Publishing Co., Ltd), February 1995.

Miyagi, M., *Fukuzatsu Daikibo System ni Okeru Jiko Boushi (II) – Koukuuki Seibi wo Megutte* [Accident Prevention in Large and Complex Systems (II) – Aircraft Maintenance] (in Japanese). Tokyo: Japan Research Institute of Air Law (sold by Yuhikaku Publishing Co., Ltd), January 1995.

Miyagi, M., *Daijiko no yochou wo saguru – Jiko e itaru michisuji wo tatsu tame ni* [Seeking Out the Signs of Major Accidents – to Break the Path that Leads to Accidents] (in Japanese). Tokyo: Kodansha Ltd, March 1998.

Other sources

AIB (UK Accident Investigation Board) Aircraft Accident Report 78, AAR-78-9.

Accident Investigation Report, Canadian Pacific Airline DC-8, CF-CPK, Tokyo International Airport, 4 March 1966; Canadian Pacific Airline and

BOAC Aircraft Accident Technical Investigation Commission (in Japanese), 4 March 1969.

Aircraft Accident Investigation Commission, Aircraft Accident Report No. 54–1 'Accident at Osaka Airport on 2 June 1978' (partially revised in June 1979) (in Japanese). Ministry of Transport Japan, February 1979.

Aircraft Accident Investigation Commission, Aircraft Accident Report No. 58–3 'Accident near Haneda Airport on 9 February 1982' (in Japanese). Ministry of Transport Japan, May 1983.

Aircraft Accident Investigation Commission, Aircraft Accident Report No. 60–5 'Accident at Naha Airport on 19 April 1984' (in Japanese). Ministry of Transport Japan, September 1985.

Aircraft Accident Investigation Commission, Aircraft Accident Report No. 60–6 'Accident at Naha Airport on 28 May 1985' (in Japanese). Ministry of Transport Japan, October 1985.

Aircraft Accident Investigation Commission, Aircraft Accident Report No. 62–2 'Accident near Osutaka Mountain on 12 August 1985' (in Japanese). Ministry of Transport Japan, June 1987.

Aircraft Accident Investigation Commission, Aircraft Accident Report No. 63–9B 'Accident at Yonago Airport on 10 January 1988' (in Japanese). Ministry of Transport Japan, October 1988.

Aircraft Accident Investigation Commission, Aircraft Accident Report No. 96–5 'Accident at Nagoya Airport on 26 April 1994' (in Japanese). Ministry of Transport Japan, July 1996.

Aircraft Accident Investigation Commission, Aircraft Accident Report No. 99–8 'Accident over Mie on 8 June 1997 (in Japanese). Ministry of Transport Japan, December 1999.

Aircraft and Railway Accident Investigation Commission, Aircraft Accident Report No. 02–5 'Accident over Yaizu on 31 January 2001' (in Japanese). Ministry of Land, Infrastructure and Transport Japan, July 2002.

All-Nippon Airways (ANA), *Koukuu Jiko Digest, Parts 1–3* (in Japanese). Edited by Comprehensive Safety Promotion Commission. Tokyo: All-Nippon Airways, 1980–81.

All-Nippon Airways (ANA), *The Montage of Aircraft Accidents I–VII* (in Japanese). Edited by Comprehensive Safety Promotion Commission. Tokyo: All-Nippon Airways, 1981.

Arseline, M. (trans. Hanaue, K.), *Airbus 320 wa naze ochita ka* [Le pilot est-il coupable?] (in Japanese). Tokyo: Kodansha, 1995.

Ashiya, S., *Kyogaku Baishou Meirei de Aratamete Towareru Keieisha Sekinin* [Management Responsibility is being Questioned Anew through Compensation Orders in Large Amounts] (in Japanese). Tokyo: Tokyo Shoko Research Ltd, 2000.

Belgium Commercial Airline Bureau, Accident Investigation Committee Report No. EW/A224, p. 25 (in English).

Benzecri, J.P., *Histoire et Préhistoire de L'Analyse des Donnée* (in French). Paris: Dunod, 1982.

Benzecri, J.P., *et al.*, *L'Analyse des Données 1, 2.* (in French). Paris: Dunod, 1973.

Benzecri, J.P. and Benzecri, F., *Pratique des L'Analyse des Donnée 1, 2, 3* (in French). Paris: Dunod, 1980.

Bundesstell für Flugunfalluntersuching, AX001-V-2/02 State Reports, 2002.

Chernobyl Ministry of Ukraine, *Ten years after the Chernobyl Accident: National Report of Ukraine* (in Russian), 1996.

'Chiezo'. *The Asahi Shimbun*, 1995.

Conner, T.M. and Hamilton, C.W., *Evaluation of Safety Programs with Report to the Cause of Air Carrier Accident*, DOT, FAA, Washington, DC, Report No. ASP80-1, January 1980.

Daikan Koukuuki Jiken no Shinsou wo Kyuumei Suru Kai [The Committee to Uncover the Truth of the Korean Air Lines Bombing] (representative: Takemono, S.), *Daikan Koukuuki Jiken no Kenkyuu* [Research into the Korean Air Lines Bombing] (in Japanese). Tokyo: San-ichi Shobo, 1987.

Diederiks-Verschoor, I.H., *An Introduction to Air Law*, 7th revised edition. The Hague: Kluwer Law International, 2001.

Domogala, P., Ruitenberg, B., and Stock, C., The mid-air collision itself: the facts. *The Controller*, Vol. 41, p. 61 onwards. IFATCA, 2002.

Eddy, P., *et al.* (trans. Ihara, T. and Kawano, K.), *Yosoku Sareta Dai-Sanji* [Major Accidents that Could Have Been Predicted] (in Japanese). Tokyo: Soshisha, 1978.

Fujita, H., *Kakusareta Shogen* [Hidden Evidence] (in Japanese). Tokyo: Shinchosha, 2003.

Fujita, T.T., *DFW Microburst on August 2, 1985.* Chicago: The University of Chicago, 1986.

Fukuda, T., *Koureisha no Shikaku Kinou* [Ageing and Visual Functions], IEICE Technical Research Report, Vol. 90, No. 409 (in Japanese), 1991.

Funatsu, Y., Koukuu no Anzen Suishin to Mondai-ten [Promoting aviation safety, and related problems] (in Japanese). *Review of Air Law and its Practice*, Vol. 15, 1983.

Gero, D. (trans. Shimizu, Y.), *Koukuu Jiko* [Aircraft Accidents] (in Japanese). Tokyo: Ikarosu Shuppan, 1994.

Goldman-Sachs, Warburg, Dylan, and Reed, *Soukaisetsu – Kinyuu Risk Management* [Comprehensive Commentary – Financial Risk Management] (in Japanese). Tokyo: The Nihon Keizai Shimbun, 1999.

Hama, R., *Yakugai wa Naze Nakunaranai ka* [Why Don't Drug-induced Disease Accidents Stop Occurring?] (in Japanese). Tokyo: Nippon Hyoronsha, 1996.

Hannson, O. (trans. Saito, M.), *Ciba Geigy no Uchimaku: Yakugai no Kouzou* [Behind the Scenes at Ciba-Geigy: The Construction of a Drug-induced Disease] (in Japanese). Tokyo: Otsuki Shoten, 1989.

Hannson, O., (trans. Yanagisawa, Y. *et al.*), *SMON Scandal: Sekai wo Mushibamu Seiyaku Gaisha* [SMON Scandal: The Pharmaceutical Company that Undermines the World] (in Japanese). Tokyo: The Asahi Shimbun, 1978.

Harada, M., *Minamata-byou* [Minamata Disease] (in Japanese). Tokyo: Iwanami Shoten, 1972.

Harada, M., *Minamata-byou kara Manabu* [Learning from Minamata Disease] (in Japanese). Tokyo: Chuohoki Publishers, 1972.

Harada, M., Bhopal: Gas Chuudoku no Koishou [Bhopal; after-effects of the gas poisoning]. *Bhopal, Shi no Toshi* [Bhopal, City of Death] (in Japanese). Tokyo: Gijutsu to Ningen [Technology and Man], 1986.

Hayashi, C., Theory and example of quantification (II) (in Japanese). Proceedings of the Institute of Statistical Mathematics 4(2): 19–30, 1956.

Hayashi, C., *Methods of Quantification* (in Japanese). Tokyo: Keizai Shinposha, 1974.

Hayashi, C., Data analysis in a comparative study. *Data Analysis and Informatics* (eds Diday, E., *et al*). The Hague: North Holland, 1980.

Hayashi, C., Quantification of qualitative data-statistical analysis of categorical data. *Analyse des Donnees, Naples, de 30 juin du 5 juillet 1980, Actes de la journee de travail.* Edited by Institut National de Recherche en Informatique et Automatique, 1980.

Hayashi, C., *Chosa no Kagaku* [The Science of Surveys] (in Japanese). Tokyo: Kodansha Blue Backs, 1984.

Hayashi, C., Statistical study on Japanese national character. *Journal of the Japanese Statistical Society (Special Issue)*: 71–95, 1987.

Hayashi, C. and Suzuki, T., Quantitative approach to a cross-societal research I: a comparative study of Japanese national character. *Annals of the Institute of Statistical Mathematics* 26: 455–516, 1974.

Hayashi, C. and Suzuki, T., Quantitative approach to a cross-societal research II: a comparative study of Japanese national character. *Annals of the Institute of Statistical Mathematics* 27: 1–32, 1975.

Hayashi, C. and Suzuki, T., Changes in belief systems, quality of life issues and social conditions over 25 years in post-war Japan. *Annals of the Institute of Statistical Mathematics* 36: 135–61, 1984.

Hayashi, C., Suzuki, T., and Hayashi, F., Comparative study of lifestyle and quality of life: Japan and France. *Behavior-metrika* 15: 1–17, 1984.

Hayashi, C., Nishihara, S., Nomoto, K., and Suzuki, T., *Hikaku Nipponjin Ron* [Comparative Studies on Japanese] (in Japanese). Tokyo: Chuokoronsha, 1973.

Heinrich, H.W., *Industrial Accident Prevention – A Scientific Approach*, 4th edition. New York: McGraw-Hill, 1931.

Heinrich, H.W., Petersen, D., and Roos, N. (trans. Sougou Anzen Kougaku Kenkyuujo [Safety Engineering Laboratories], Supervising Ed. Inoue, T.),

Heinrich to Sangyou Saigai Boushi-ron [Industrial Accident Prevention: Safety Management Approach] (in Japanese). Tokyo: Kaibundo, 1982.

Hirokawa, R., *Yakugai AIDS* [Drug-induced AIDS] (in Japanese). Tokyo: Iwanami Shoten, 1995.

Hirono, R., *Sabakareru Yakugai Aids* [Judgement on Drug-induced AIDS] (in Japanese). Tokyo: Iwanami Shoten, 1996.

Hiwatari, J. and Ashida, H., *Operational Risk Kanrino Koudo-ka ni kansuru ronten seiri to kongo no kadai – teigi-teki risk kanri shouhou dounyuu e no torikumi wo chuushin ni* [Summary of Key Points and Future Issues Related to the Improvement of Operational Risk Management – Focusing on Efforts Targeting the Introduction of Defined Risk Management Methods] (in Japanese). Tokyo: Bank of Japan Assessment Bureau, Discussion Paper No. 02-J-1, 2002.

International Civil Aviation Organization (ICAO) Accident Digest, Circular 47-AN/42 (16–45), Report of the public inquiry into the causes and circumstances of the accident that occurred on the 10 January 1954 to the Comet aircraft G-ALYP.

ICAO Accident/Incident Data Reporting.

ICAO Accident Prevention Manual, Doc 9422-AN/923, 1984.

ICAO Aircraft Accident Digest, Circular 153–AN/56. *KLM B-747, PH-BUF and Pan Am B-747 N736 Collision at Tenerife Airport Spain on 27 March 1977*, Report dated October 1978, released by the Subsecretaria de Aviacion Civil, Spain.

ICAO Annual Report of the Council, 1983.

ICAO Digest No. 1, CAP 719.

ICAO Fact-Finding Investigation C-WP/7764, December 1983, *Destruction of Korean Air Lines Boeing 747 Over Sea of Japan, 31 August 1983*.

ICAO Human Factors Training Manual, Doc 9683-AN/950, 1998.

Iguchi, T., *Kokuhaku* [Confession] (in Japanese). Tokyo: Bungeishunju, 1999.

Ikeda, E., *AIDS to Ikiru Jidai* [The Era of Living with AIDS] (in Japanese). Tokyo: Iwanami Shoten, 1993.

Ikeda, F., *Shiroi Ketsueki – AIDS Kansen to Nihon no Ketsueki Sangyo* [White Blood – AIDS Infection and Japan's Blood Industry] (expanded edition, in Japanese). Tokyo: Ushio Shuppansha, 1985.

Imanaka, T., *Kokusai Kyoudou Kenkyuu Houkokusho: Chernobyl Jiko ni yoru Houshanou Saigai* [International Joint Research Report: Radioactive Damage Resulting from the Chernobyl Accident] (in Japanese). Tokyo: Gijutsu to Ningen [Technology and Man], 1998.

Iryou Jiko Chousakai [Medical Accident Investigation Committee, Ed.], *Iryou jiko wo fusegu tame ni – Symposium kiroku shuu* [To Prevent Medical Accidents – Symposium Notes] (in Japanese). Tokyo: Medical Tribune Inc., 1998.

Ishida, Y. and Murakami, H., *AIDS to Ikiru* [Living with AIDS] (in Japanese). Tokyo: Hosei University Publishing Office, 1994.

Ito, H. (Ed.), *Koukuu Kishou* [Aviation Meteorology], 7th edition (in Japanese). Tokyo: Tokyodo Shuppan, 1986.

Iwahara, S., *Daiwa Ginkou Daihyou Soshou Jiken Isshin Hanketsu to Daihyou Soshou Seido Kaisei Mondai* [The First Judgement in the Daiwa Bank Representatives Lawsuit and the Problem of Improving the Representative Lawsuit System] (in Japanese). Tokyo: Shouji Houmu, Nos 1576 and 1577, 2001.

JAL Flight Safety, *Berugii Minkan Koukuukyoku Jiko Chousa Iinkai Jiko Houkokusho* [Belgian Commercial Airline Bureau, Accident Investigation Committee Report No. EW/A224; abridged translation], No. 59 (in Japanese), 1988.

JAL Flight Safety, No. 68, Testimony by Dr Colen Drury, February 1990.

Japan Aeronautical Engineer's Association, *Aviation Engineering*, No. 412 (in Japanese), 1989.

Japan Aircraft Pilot Association, *Pilot*, No. 5 (in Japanese), 1985.

Japan Flight Crew Union Federation, *Nikkou 123 bin ni kyuu gen'atsu wa nakatta* [There Was No Rapid Decompression on JAL Flight 123] (in Japanese), 1994.

Japan Research Institute of Air Law, Koukuuki Jiko Boushi no Tame no Teigen – Jiko Chousa wo chuushin toshite [Proposals for preventing aviation accidents – with a focus on accident investigations], *Review of Air Law and its Practice* (in Japanese), Vol. 13, 1980.

Japan Research Institute of Air Law, Koukuuki Jiko Boushi no Tame no Teigen – Total Air Safety System kenkyuu no yousei [Proposals for Preventing Aviation Accidents – Requests for research in Total Air Safety Systems], *Review of Air Law and its Practice* (in Japanese), Vol. 15, 1983.

Job, M., *Air Disaster,* Vol. 1. Australia: Aerospace Publications, 1997.

Kakuta, S., *Giwaku – JAL 123 bin tsuiraku jiko* [Doubt – JAL123 Crash Accident] (in Japanese). Tokyo: Waseda Shuppan, 1993.

Kato, K., *Kowareta Biyoku* [Broken Tail Plane] (in Japanese). Tokyo: Gihodo Shuppan, 1987.

Kato, K., *Tsuiraku* [Airplane Crashes] (in Japanese). Tokyo: Kodansha, 1990.

Kato, K., *Kanseikan no Ketsudan (Near Miss)* [Decisions of Air Traffic Controllers (Near Misses)] (in Japanese). Tokyo: Kodansha, 1992.

Kawaguchi, Y., *Hanketsu 'Heisei 12 nendo Jyuuyou Hanrei Kaisetsu'* [Judgement 'Explanation of Important Precedents in 2000'] (in Japanese). Tokyo: Jurist, No. 1202, 2001.

Kawamoto, I., Daiwa Ginkou Kabunushi Daihyou Soshou no Wakai [Settlement in the Daiwa Bank Shareholders Representatives Lawsuit] (in Japanese). *Torishimariyaku no Houmu* [The Stance of Directors], No. 94, p. 44, 2002.

Kawamura, M., *Hanketsu* [Judgement] (in Japanese). Tokyo: Kinyu/Shouji Hanrei [Financial/Commercial Precedents], No. 1107, 2001.

Kawasaki, T., *America Kinyu Hanzai no Ichi Danmen – Daiwa Ginko NY Shiten*

Jiken kara no Kyokun [A Cross-section of American Financial Crime – Lessons from the Daiwa Bank NY Branch Scandal] (in Japanese). Tokyo: Shouji Houmu, No. 1602, p. 51, 2001.

Kikuchi, S., *Daiwa Ginkou Kabunushi Daihyou Soshou Hanketsu ni Furete* [Touching on the Decision in the Daiwa Bank Shareholders Representatives Lawsuit] (in Japanese). Tokyo: Kansayaku [Auditor], No. 436, p. 9, 2001.

Kondo, J., *Bhopal – Shijou Saiaku no Kagaku Saigai* [Bhopal – The Worst Chemical Disaster in History] (in Japanese). Tokyo: Gijutsu to Ningen [Technology and Man], 1985.

Kondo, J., *Kyodai System no Anzensei* [Safety in Large-scale Systems] (in Japanese). Tokyo: Kodansha, 1987.

Konoplya, E.F. and Rolovich, I.V. (Eds), *Ecological, Medicobiological and Socioeconomic Consequences of the Catastrophe at the Chernobyl NPS in Belarus* (in Russian). Ministry of Emergency and Chernobyl Problems of Belarus, Institute of Radiobiology of Academy of Sciences of Belarus, 1996.

Kyoto University KURRI-KR-21, *Research Activities about the Radiological Consequences of the Chernobyl NPS Accident and Social Activities to Assist the Sufferers by the Accident*, 1998.

Leeson, N. (trans. Toda, H.), *Watashi ga Barings Ginko wo Tsubushita* [Rogue Trader] (in Japanese). Tokyo: Shinchosha, 1997.

MacPherson, M. (trans. Yamamoto, M.), *Tsuiraku no Shunkan – Voice Recorder ga kataru Shinjitsu* [The Black Box] (in Japanese). Tokyo: Aoyama Publishing, 1999.

Medvejef, J.A. (trans. Umebayashi, H.), *Ural no Kaku Sanji* [Nuclear Disaster in the Urals], 6th edition (in Japanese). Tokyo: Gijutsu to Ningen [Technology and Man], 1988.

Medvejef, J.A. (trans.Yoshimoto, S.), *Chernobyl no Isan* [Chernobyl's Legacy] (in Japanese). Tokyo: Misuzu Shobo, 1992.

Ministry of Health and Welfare SMON Research Group, Reports (various years).

Ministry of Transport Japan, '*Koukuu Unyu Toukei Nenpyou*' Koukuu Seisaku *Kenkyuukai Kouseiken Series* ['Air Transport Statistics Annual Report' Air Measures Research Institute Kouseiken Series], No. 197, material 12 (in Japanese).

Ministry of Transport Japan, Civil Aviation Bureau, 5 August 1970, Aircraft Investigation No. 68 (in Japanese).

Minutes of the 101st National Assembly House of Councillors Committee on Budget (23 February 1984).

Minutes of the 112th National Assembly House of Councillors Committee on Audit (25 April 1988).

Miyagi, M., Koukuu no Anzen Kakuho to Jouhou Koukai [Assuring aviation safety and the disclosure of information]. *Jouhou Koukai to Gendai* [Information Disclosure and Today], p. 108 onwards (in Japanese). Tokyo: Nippon Hyoronsha, 1982.

Miyagi, M., Anzen Houkoku Seido (IRS) no Kakuritsu Yousei to Mondaiten – Koukuu Anzen no Koujou [Request for the establishment of a safety reporting system (IRS) and related problems – improving aviation safety]. *Yobo Jiho*, No. 147 (in Japanese). Tokyo: The General Insurance Organization of Japan, 1986.

Miyagi, M., Himitsu wa giwaku no kongen – Koukuuki Jiko Chousa to Anzen no Kangaekata [Secrecy is the root of suspicion – aircraft accident investigations and approaches to safety]. *Sekai* [World] (in Japanese). Tokyo: Iwanami-shoten, October 1985.

Miyagi, M., Fukuzatsu Daikibo System ni Okeru Jiko no Mizen Boushi no Hitsuyousei – Koukouki Seibi wo Megutte [The need for prior prevention of accidents in complex large-scale systems – focusing on aircraft maintenance]. *The 26th Safety Engineering Symposium*, p. 133 (in Japanese). Tokyo: Science Council of Japan, Safety Engineering Research Commission, 1986.

Miyagi, M., Incident Reporting System no Sangyou e no Ouyou [Applying incident reporting systems to various fields of industry]. *Safety* (in Japanese). Tokyo: Japan Industrial Safety and Health Association, 1987.

Miyagi, M., *Koudo Gijutsu Shakai ni Okeru Anzen Kanri System* [Safety Administration Systems in Advanced Technological Societies] (in Japanese). Tokyo: The Toyota Foundation, 1987.

Miyagi, M., Koukuuki Soujuu-jou no Human Factors to Anzen no Kakuho – IRS no Hitsuyousei [Human factors in the operation of aircraft and the assurance of safety – the need for IRS]. *SUT Bulletin* (in Japanese). Tokyo: Science University of Tokyo Publishing Office, 1987.

Miyagi, M., Jinteki Youin Incident no Bunseki Shuhou no Kaihatsu to sono Kyakkanteki Datousei no Jisshou [Development of analysis methods in human factor incidents and the objective verification of appropriateness]. *Proceedings of the 26th Aircraft Symposium 1988*, p. 640 (in Japanese). The Japan Society for Aeronautical and Space Sciences, 1988.

Miyagi, M., Koukuu ni Okeru IRS no Hitsuyousei [The need for IRS in aviation]. *Proceedings of the 26th Aircraft Symposium 1988*, p. 639 (in Japanese). The Japan Society for Aeronautical and Space Sciences, 1988.

Miyagi, M., Koukuuki nado Kyodai System ni Okeru Senzaiteki Kiken Youin no Hakkutsu hou [New method of discovering the latent dangerous factors in large-scale systems, air transportation, etc.]. RC83 Research Report (in Japanese). Tokyo: The Japan Society of Mechanical Engineers, 1989.

Miyagi, M., Koukuu ni Okeru Jinteki Youin Incident to Jiko no Mizen Boushi [Preventing aircraft accidents in advance by analysis of human factor incidents involved in the aircraft operations]. *13th Symposium on Industrial Safety Measures*, Vols 8-2-1 to 8-2-9 (in Japanese). Tokyo: Japan Management Association, 1990.

Miyagi, M., *Koukuuki Unkou ni Okeru Jinteki Youin – Tahenryou Kaiseki no Ouyou* [Human Factors in Aircraft Operations – Applications of multivariate analysis] (in Japanese). Tokyo: The Operations Research Society of Japan, 1990.

Miyagi, M., *Nihon Minkan Teiki Koukuu ni okeru Senzaiteki Kiken Youin no Nidai Tokusei* [The Two Major Characteristics of the Latent Dangerous Factors in the Japanese Civil Air Service Operation] (in Japanese). Tokyo: Reliability Engineering Association of Japan, 1990.

Miyagi, M., Development of MAIR, a Method for Analyzing Human Factor Incidents and the Need for an Integrated Multi-Dimensional Approach – Based on Incident Reports by Airline Pilots. *Proceedings of the First Beijing International Conference on Reliability, Maintainability and Safety (BICRMS '92)*, p. 239 onwards. Beijing: International Academic Publishers, 1992.

Miyagi, M., Jiko Chousa no Mondai ni tsuite – Koukuuki Seibi no Incident o Megutte [The problems of accident investigations – focusing on aircraft maintenance incidents]. Research Report No. 15 and Security, No. 82 (in Japanese). Tokyo: Secom Science and Technology Foundation, and Security World, 1996.

Miyagi, M., Jinteki Youin Incident no Shin Bunseki Shuhou – MAIR to Tajigenteki Kousatsu no Hitsyousei [Development of MAIR a method for analysing human factor incidents and the need for an integrated multi-dimensional approach]. *Reliability Engineering Association of Japan Journal – Reliability*, Vol. 19, No. 2, pp. 10–20 (in Japanese). Tokyo: Reliability Engineering Association of Japan, 1997.

Miyagi, M., Fuanzen Jishou no Tokushitsu to Sougouteki / Tajigenteki Kousatsu [Comprehensive and multi-dimensional investigations of unsafe events]. Technical report of IEICE, pp. 39–44 (in Japanese). Tokyo: The Institute of Electronics, Information and Communication Engineers, 2001.

Miyagi, M., *IRAS no Jisshou* [Demonstration of IRAS] (in Japanese). Ningen Kogakukai Koukuu Ningen Kougakubukai [Japan Ergonomics Society – Aviation Human Factors Division], 2002.

Miyamoto, M., *Thalidomide Wazawai no Hitobito* [People of the Thalidomide Disaster] (in Japanese). Tokyo: Chikuma Shobou, 1981.

Miyao, K., Hirou to Shikinou [Fatigue and visual function]. *Igaku to Kougaku kara mita Koutsuu Anzen Taisaku* [Traffic Safety Measures as Seen from Medicine and Engineering] (in Japanese), 1994.

Mizuno, T., *New York Hatsu Daiwa Ginkou Jiken* [The Daiwa Bank Scandal from New York] (in Japanese). Tokyo: Diamond Inc., 1996.

Moore-Ede, M. (trans. Aoki, K.), *Daijiko wa yoake-mae ni okiru* [Major Accidents Happen Before Dawn] (in Japanese). Tokyo: Kodansha, 1994.

Mori, M. (Ed.), *Nigai Karute – Shourei ni Manabu Goshin Yobougaku* [Bitter Charts – Misdiagnosis Prevention as Learned from Actual Cases] (in Japanese). Tokyo: Nikkei Mcgraw-Hill, 1984.

Muranaka, J., *Pilot no tachiba kara mita mondai-ten to jitsujou* [Problems and

Actual Conditions from the Pilot's Perspective] (in Japanese). *Review of Air Law and its Practice*, Vol. 15, 1983.

Nagasu, H., Koukuu Anzen no Tame no Gijutsu-teki Sho-mondai [Technical problems in achieving aviation safety] (in Japanese). *Review of Air Law and its Practice*, Vol. 15, 1983.

Nakamura, N., *Daiwa Ginkou Jiken Hanketsu to Daihyou Soshou no Arikata* [The Daiwa Bank Scandal Judgement and the Form of Representative Lawsuits] (in Japanese). Tokyo: Jurist, No. 1191, p. 20, 2001.

Nanasawa, K, *Genpatsu Jiko wo Tou* [Inquiring into Nuclear Accidents] (in Japanese). Tokyo: Iwanami Shoten, 1996.

National Academy of Sciences, *Low Altitude Wind Shear and its Hazard to Aviation*, pp. 14–15. Washington, DC: The National Academic Press, 1983.

National Research Council, *Aviation Safety and Pilot Control*. Washington: National Academic Press, 1997.

National Transportation Safety Board (NTSB) Report Number AAR-72-12, *Capitol International Airways, DC-8-63F, N4909C, Anchorage, Alaska, November 27, 1970*. Adopted on 29 March 1972.

NTSB Report Number AAR-73-14, *Eastern Air Lines, Inc., L-1011, N310EA, Miami, Florida, December 29, 1972*. Adopted on 14 June 1973.

NTSB Report Number AAR-73-15, *North Central Airlines, Inc., McDonnell Douglas DC-9-31, N954N and Delta Air Lines, Inc., Convair CV-880, N8807E, O'Hare Int'l Arpt. Chicago, IL, December 20, 1972*. Adopted on 5 July 1973.

NTSB Report Number AAR-75-09, *Eastern Air Lines, Inc., Douglas DC-9-31, N8984E, Charlotte, North Carolina, September 11, 1974*. Adopted on 23 May 1975.

NTSB Report Number AAR-75-16, *Trans World Airlines, Inc., Boeing 727-231, N54328, Berryville, Virginia, December 1, 1974*. Adopted on 26 November 1975.

NTSB Report Number AAR-76-08, *Eastern Airlines, Inc., Boeing 727- 225 JFK International Airport, Jamaica, New York, June 24, 1975*. Adopted on 12 March 1976.

NTSB Report Number AAR-76-12, *Japan Air Lines, Company, Ltd, Boeing 747-246, JA8122, Anchorage, Alaska, December 16, 1975*. Adopted on 31 March 1976.

NTSB Report Number AAR-76-15, *Eastern Air Lines, Inc., Boeing 727-225, N8838E, Raleigh, North Carolina, November 12, 1975*. Adopted on 19 May 1976.

NTSB Report Number AAR-79-12, *Allegheny Airlines, Inc., Nord 262 Mohawk/Frakes 298, N29824-Benedum Airport, Clarksburg, West Virginia, 12 February 1979*. Adopted on 16 August 1979.

NTSB Report Number AAR-79-17, *American Airlines, Inc., DC-10, N110AA, Chicago International Airport, Chicago, IL, May 25, 1979*. Adopted on 21 December 1979.

NTSB Report Number AAR-82-08, *Air Florida, Inc., Boeing 737-222, N62AF, Collision with 14th Street Bridge, near Washington Nat'l Airport, Washington, DC, January 13, 1982.* Adopted on 10 August 1982.

NTSB Report Number AAR-89-03, *Aloha Airlines, Flight 243, Boeing 737-200, N73711, Near Maui, Hawaii, April 28, 1988.* Adopted on 14 June 1989.

NTSB Report Number AAR-90-01, *United Airlines Flight 811 Boeing 747-122, N4713U Honolulu, Hawaii, February 24, 1989.* Adopted on 16 April 1990.

NTSB Report Number AAR-91-04, *Avianca, The Airline of Columbia, Boeing 707-321B, HK 2016, Fuel Exhaustion, Cove Neck, New York, January 25, 1990.* Adopted on 30 April 1991.

NTSB Report Number AAR-91-05, *NW Airlines, Inc., Flights 1482 and 299 Runway Incursion and Collision, Detroit Metropolitan/Wayne County Airport, Romulus, Michigan, December 3, 1990.* Adopted on 25 June 1991.

NTSB Report Number AAR-92-02, *Explosive Decompression – Loss of Cargo Door in Flight, United Airlines Flight 811 Boeing 747-122, N4713U Honolulu, Hawaii, February 24, 1989. Revised.* Adopted on 18 March 1992.

NTSB Identification DCA77RA014, 14 CFR General Aviation Form. Event occurred Sunday 27 March 1977 in Tenerife, Canr, Spain. Aircraft: Boeing 747, registration PH-BUP.

NTSB Identification MIA94FA169, 14 CFR 121 operation of American Airlines, Inc. Accident occurred Wednesday 29 June 1994 in Caribbean, CB. Aircraft: McDonnell Douglas MD-11, registration N1752K.

NTSB Identification SEA69A0062, 14 CFR General Aviation Form. Event occurred Tuesday 24 June 1969 in Moses Lake, WA. Aircraft: Convair 880, registration JA-8028.

Newhouse, J. (trans. Aircraft Industry Research Group), *Sporty Game*, edited by Ishikawajima-Harima Heavy Industries Co., Ltd (in Japanese). Tokyo: Gakuseisha, 1988.

Nishida, S., Gansoshiki no rouka to chousetsu [Ageing changes of ocular tissues and their influences on accommodative functions] (in Japanese). *Japanese Ophthalmological Society Journal*, Vol. 94, No. 2, 1989.

Okinaka Shigeo Sensei wo Shinobu Kai [Published in memory of Dr Shigeo Okinaka]. *Okinaka Shigeo – I no michi* [*Shigeo Okinaka – the Way of Medicine*] (in Japanese). Tokyo: Japan Medical Journal, 1992.

Osaka HIV Lawsuit Legal Team (Ed.), *Yakugai AIDS Kokusai Kaigi* [International Meeting on Drug-induced AIDS] (in Japanese). Tokyo: Sairyuusha, 1998.

Oshima, M., Medical check no mondai-ten to taisaku [Medical checks: problems and countermeasures]. *Review of Air Law and its Practice*, Vol. 15, p. 79 onwards (in Japanese). Tokyo: Japan Research Institute of Air Law (sold by Yuhikaku Publishing Co., Ltd), 1983.

Oshima, M., Yamamoto, S., and Yokobori, S., *Koukuu Igaku* [Airline Medicine] (in Japanese). Tokyo: Igaku Shoin, 1967.

Owen, D. (trans. Aoki, Y.), *Tsuiraku Jiko – Kitai ga Kataru Tsuiraku no*

Scenario [Airplane Crashes – the Airplane Body Tells the Story of the Crash Scenario] (in Japanese). Tokyo: Hara Shobo, 2003.

Power Reactor and Nuclear Fuel Development Corporation, *Monju Natoriumu Rouei Jiko no Gaiyou* [Outline of the Monju Sodium Leak Accident] (in Japanese).

Report de la commission d'enquête sur l'accident survenu le 20 janvier 1992 près de Mont Sainte-Odile (Bas Rhin) à l'Airbus A320 immatriculé F-GGED exploité par la compagnie Air Inter F-ED 920/120 (in French).

Report of the Court Investigation India, Accident to Air India Boeing 747 aircraft VT-EFO, 'KANISHKA', on 23 June 1985, 1986.

Report of the President's Commission on the Three Mile Island Accident (in English). *Three-mile tou genpatsu Jiko Houkokusho* [Three Mile Island Nuclear Accident Report] (in Japanese). Tokyo: Highlife Shuppan, 1977.

Republic of South Africa, Report of the Board of Inquiry into the Helderberg Air Disaster, ISBN0-621-13030-3 Appendix A.

Sakai, I., Kuni niyoru Incident Reporting System [Incident reporting systems on the national level]. *Review of Air Law and its Practice*, No. 15, p. 35 onwards (in Japanese), 1983.

Sakamoto, T., *Atarashii Kokusai Koukuu-hou* [New International Aviation Law] (in Japanese). Tokyo: Yushindo, 1999.

Sakamoto, T., *Gendai Kouku-hou* [Modern Aviation Law] (in Japanese). Tokyo: Yushindo, 1984.

Sakamoto, T., *Gendai Kuu-unyu* [Modern Air Transport] (in Japanese). Tokyo: Seizando, 1988.

Sakamoto, T., *Kokusai Koukuu Hou Ron* [International Aviation Law Theory] (in Japanese). Tokyo: Yushindo, 1992.

Sakamoto, T., *Yomigaere Nihon no Tsubasa* [Come Back to Life, Japan's Wings] (in Japanese). Tokyo: Yushindo, 2003.

Sakurai, H., *Umoreta AIDS Houkoku* [The Buried AIDS Report] (in Japanese). Edited by NHK Reporting Group. Tokyo: Sanseido, 1997.

Sasada N., Kakukoku no Incident Reporting System nituite [Incident reporting systems in various countries and states]. *Review of Air Law and its Practice*, 1981, No. 13, p. 37 onwards and No. 14, p. 5 onwards (in Japanese). Tokyo: Yuhikaku Publishing Co., Ltd, 1981.

Schelbak, Y. (trans. Matsuoka, N.), *Chernobyl kara no shougen* [Testimonies from Chernobyl] (in Japanese). Tokyo: Gijutsu to Ningen [Technology and Man], 1988.

Schiavo, M. (trans. Sugiura, K., et al.), *Abunai Hikouki ga Kyou mo Tonde-iru* [Dangerous Airplanes are Flying Today, Too] (in Japanese). Tokyo: Soshisha, 1999.

Science and Technology Agency, *Kousoku Zoushoku Genkeiro Monju Natoriumu Rouei Jiko no Houkoku* [Report on Sodium Leak Accident at the Power Reactor and Nuclear Fuel Development Corporation's Monju Prototype Fast-Breeder Reactor] (in Japanese), 1998.

Serling, R.J. (trans. Fukushima, M.), *Koukuu Anzen Kakumei* [Aviation Safety Revolution] (in Japanese). Tokyo: Diamond, Inc., 1971.

Shibata, T., Dai san-sha no Tachiba kara mita Mondai no Shozai to Jouhou Koukai no Hitsuyousei [Location of problems from a third-party perspective, and the need for information disclosure] (in Japanese). *Review of Air Law and its Practice*, Vol. 15, 1983.

Shibata, T., *Kagaku Houdou* [Science Reporting] (in Japanese). Tokyo: Asahi Shimbun, 1994.

Shimizu, Y., *Thatcher and Shultz* (in Japanese). Tokyo: Josuikai New Year's Edition, 2003.

Shiomi, H., *Jiko wa naze nakunaranai ka* [Why Don't Accidents Stop Occurring?] (in Japanese). Tokyo: The General Insurance Organization of Japan, 1985.

Shiomi, H., *Ningen Shinraisei Kougaku Nyuumon* [Introduction to Human Reliability Engineering] (in Japanese). Tokyo: Union of Japanese Scientists and Engineers (JUSE), Publishing Division, 1996.

Silverman, M.M. (trans. Saito, M.), *Iyakuhin Scandal – Daisan Sekai e no Iyakuhin Dumping* [Bad Medicine: The Prescription Drug Industry in the Third World] (in Japanese). Tokyo: San-ichi Shobo, 1986.

Subsecretaria de Aviation Civil, Spain, released, Part 1, Comments of the Netherlands Department of Civil Aviation, Report dated October 1978.

Swain, N.K. (trans. Hisatome, K.), Ima, Higaishatachi wa [Where are the victims now?]. *Bhopal, Shi no Toshi* [Bhopal, City of Death] (in Japanese). Tokyo: Gijutsu to Ningen [Technology and Man], 1986.

Takagi, J., *Kyodai Jiko no Jidai* [The Era of Major Accidents] (in Japanese). Tokyo: Koubundou, 1988.

Taketani, M. (Ed.), *Anzensei no Kangaekata* [Approaches to Safety] (in Japanese). Tokyo: Iwanami Shoten, 1967.

Takeuchi, E., *Genshiryoku Shisetsu ni Okeru Safety Culture no Jousei ni tsuite – sono Kokoro* [Cultivating a Safety Culture in Atomic Power Facilities – The Heart] (in Japanese). Japan Electric Association, Newspaper Division, 1997.

The Observer (Ed.) (trans. Kawanago, M., Sawa, H., and Yamada, S.), *Shijou Saiaku no Kaku Osen* [The Worst Radiation Pollution in History] (in Japanese). Tokyo: Sankei Shuppan, 1986.

The Yomiuri Shimbun, Science Division, *Document Monju Jiko* [Documented Monju Accident] (in Japanese). Tokyo: Million Publishing, 1996.

Tokoro, T., Gan Kinou no Nenrei Henka [Changes in visual functions due to age]. *Shiryoku zukai rinshou ganka kouza: roujin to me* [Vision: an Illustrated Course on Clinical Opthamology: the Elderly and Eyes] (in Japanese). Tokyo: Medical Review Co., Ltd, 1986.

Tsuruoka, K., and Kitamura, Y., *Higeki no Shinsou – Nikkou Jumbo-ki Jikou Chousa no 667 nichi* [The Truth of the Tragedy – 667 days in the Investigation of the JAL Jumbo Jet Accident] (in Japanese). Tokyo: The Yomiuri Shimbun, 1991.

Ueda, T., Raiun no Katsudo [Activities of thunderclouds]. *Aviation Engineering*, No. 396 (in Japanese). Tokyo: Japan Aeronautical Engineer's Association, 1988.

Webb, R.E., The health consequences of Chernobyl. *The Ecologist*, 16, 1986.

West Research Council, *Aviation Safety and Pilot Control – Understanding and Preventing Unfavorable Pilot – Vehicle Interactions*. Washington: National Academic Press, 1977.

Winner, E.L., *Human Factors in Aviation* (Ed. Nagel, D.C.). San Diego: Academic Press, Inc., 1988.

World Watch Institute, Brown, L. R. (Ed.) (chief trans. Honda, Y.), *Chikyuu Hakusho* [Earth White Paper] (in Japanese). Tokyo: Diamond Inc., 1988.

Yamaguchi, M., *Koukuu Jiko Chousa – Seido to Unyou* [Aircraft Accident Investigations – Systems and Operation] (in Japanese). Tokyo: Hobun Shorin, 1979.

Yamana, M. *Saigo no 30-byo* [The Last 30 Seconds] (in Japanese). Tokyo: Asahi Shimbun, 1975.

Yanagida, K., *Jiko Chousa* [Accident Investigations] (in Japanese). Tokyo: Shinchosha, 1994.

Yanagida, K., *Koukuu Jiko* [Aircraft Accidents] (in Japanese). Tokyo: Chuko Shinsho, 1975.

Yasuda Kasai Kaijou – Kankyou Kouza [Yasuda Fire and Marine Insurance Company, Limited – Environment Course] (in Japanese). Tokyo: Chouhoki, 1997.

Yoshikawa, T. and Kimura, F., *Chernobyl Genpatsu Jiko ni yoru Houshanousei Busshitsu no Chikyuu Kibo Kakusan* [Global Scale Dispersal of Radioactive Materials due to the Chernobyl Nuclear Accident] (in Japanese). Tokyo: Kishou Shunju, Vol. 73, 1987.

Notes on the text

Chapter 1

1 Takagi, J. (1989), *Kyodai Jiko no Jidai* [The Era of Major Accidents] (in Japanese). Tokyo: Koubundou, p. 143.

 All conversions from roubles to US dollars in this text are based on exchange rates at the end of June 2003 ($1 = 30.30 roubles).

2 Imanaka, T. (1998), *Kokusai Kyoudou Kenkyuu Houkokusho: Chernobyl Jiko ni yoru Houshanou Saigai* [International Joint Research Report: Radioactive Damage Resulting from the Chernobyl Accident]. Tokyo: Gijutsu to Ningen [Technology and Man] (in Japanese), p. 365.

 Research Activities about the Radiological Consequences of the Chernobyl NPS Accident and Social Activities to Assist the Sufferers by the Accident (1998). Kyoto University KURRI-KR-21 (in English).

 Konoplya, E. F. and Rolovich, I.V. (Eds), *Ecological, Medicobiological and Socioeconomic Consequences of the Catastrophe at the Chernobyl NPS in Belarus*. Ministry of Emergency and Chernobyl Problems of Belarus, Institute of Radiobiology of Academy of Sciences of Belarus, 1996 (in Russian).

 All expenses related to Chernobyl accident measures up to September 1991 were paid out from the budget of the former Soviet Union. After this time, funds were paid from the national budget of the Ukraine through the Chernobyl Fund. Chernobyl Ministry of Ukraine, *Ten Years after the Chernobyl Accident: National Report of Ukraine*, 1996 (in Russian).

 According to Nanasawa, K. (1996), *Genpatsu Jiko wo Tou* [Inquiring into Nuclear Accidents] (in Japanese), Tokyo: Iwanami Shoten, p. 230, the amount paid out by the Soviet Union from 1986 to 1991 is said to be 25 billion roubles (825 million dollars).

3 According to the World Watch Institute, the United Kingdom, Sweden, West Germany, and Poland are said to have estimated damages at a total of 450 million dollars. {Brown, L. R. (Ed.) (chief trans. Honda, Y.) (1988), *Chikyuu Hakusho* [Earth White Paper] (in Japanese). Tokyo: Diamond Inc}.

4 There are varying reports on the number of deaths. An official report from

the state government placed the number at 1754, while some media reported 3500 deaths. Most media put the number at 2000.

5 Swain, N. K. (trans. Hisatome, K.), Ima, Higaishatachi wa [Where are the victims now?]. *Bhopal, Shi no Toshi* [Bhopal, City of Death]. Tokyo: Gijutsu to Ningen [Technology and Man] (in Japanese), p. 21.

   Kondo, J. (1986), *Kyodai System no Anzensei* [Safety in Large-scale Systems] (in Japanese). Tokyo: Kodansha, p. 26.

6 Harada, M. (1986), Bhopal: Gas Chuudoku no Koishou [Bhopal: After-effects of the Gas Poisoning]. *Bhopal, Shi no Toshi* [Bhopal, City of Death]. Tokyo: Gijutsu to Ningen [Technology and Man] (in Japanese), p. 120 onwards.

7 Kondo, J. (1986), *Kyodai System no Anzensei* [Safety in Large-scale Systems] (in Japanese). Tokyo: Kodansha, p. 26.

8 *Ibid.*, p. 34.

9 *Ibid.*, p. 42.

   The Yomiuri Shimbun, 23 November 1986.

10 Kondo, J. (1986), *Kyodai System no Anzensei* [Safety in Large-scale Systems] (in Japanese). Tokyo: Kodansha, p. 143.

   Decisive evidence of the occurrence of this accident was provided by a photo of the roof of the power plant being blown off by the force of the explosion, taken by the American reconnaissance satellite KH11.

11 *The Mainichi Shimbun*, 12 May 1986, evening edition.

   According to Shimizu, Y. (2003), *Thatcher and Shultz.* Tokyo: Josuikai, 2003, New Year's Edition, p. 6 (in Japanese), Gorbachev first learned of this accident from a BBC broadcast.

12 Science and Technology Agency (1998), *Kousoku Zoushoku Genkeiro Monju Natoriumu Rouei Jiko no Houkoku* [Report on Sodium Leak Accident at the Power Reactor and Nuclear Fuel Development Corporation's Monju Prototype Fast-breeder Reactor] (in Japanese).

13 Nanasawa, K. (1996), *Genpatsu Jiko wo Tou* [Inquiring into Nuclear Accidents] (in Japanese). Tokyo: Iwanami Shinsho, p. 10.

   Power Reactor and Nuclear Fuel Development Corporation, *Monju Natoriumu Rouei Jiko no Gaiyou* [Outline of the Monju Sodium Leak Accident] (in Japanese), p. 3.

   The Yomiuri Shimbun, Science Division (1996), *Document Monju Jiko* [Documented Monju Accident] (in Japanese). Tokyo: Million Publishing, p. 54 onwards.

   The Mainichi Shimbun, 21 December 1995, morning edition and evening edition.

   The Tokyo Shimbun, 21 December 1995.

   The Asahi Shimbun, 25 December 1995.

14 *The Denki Shimbun*, 18 April 1997.

   The Nihon Keizai Shimbun, 17 April 1997.

15 Medvejef, J. A. (trans. Umebayashi, H.), 6th edition (1988), *Ural no Kaku Sanji* [Nuclear Disaster in the Urals]. Tokyo: Gijutsu to Ningen [Technology and Man] (in Japanese), p. 95.

16 Medvejef, J. A. (trans. Yoshimoto, S.) (1992), *Chernobyl no Isan* [Chernobyl's Legacy] (in Japanese). Tokyo: Misuzu Shobo, p. 306.

In January 1980, there was an accident at the Kursk nuclear power plant, which was equipped with a reactor, but, in this accident, power was shut down completely so there were no major problems while the reactor was operational (*ibid.*, p. 298).

17 Schelbak, Y. (trans. Matsuoka N.) (1988), *Chernobyl kara no shougen* [Testimonies from Chernobyl]. Tokyo: Gijutsu to Ningen [Technology and Man] (in Japanese), p. 29 onwards. This report was published 27 March 1986, in the magazine *Literaturna Ukraine* under the title 'No Special Problem').

18 Takagi, J. (1989), *Kyodai Jiko no Jidai* [The Era of Major Accidents] (in Japanese). Tokyo: Koubundou, p. 150.

19 The Observer (Ed.) (trans. Kawanago, M., Sawa, H., and Yamada, S.) (1986), *Shijou Saiaku no Kaku Osen* [The Worst Radiation Pollution in History] (in Japanese). Tokyo: Sankei Shuppan, p. 134 onwards.

20 Yoshikawa, T. and Kimura, F. (1987), Chernobyl Genpatsu Jiko ni yoru Houshanousei Busshitsu no Chikyuu Kibo Kakusan [Global scale dispersal of radioactive materials due to the Chernobyl nuclear accident] (in Japanese). *Kishou Shunju*, Vol. 73 (20 March 1987), p. 6 onwards.

21 Takagi, J. (1989), *Kyodai Jiko no Jidai* [The Era of Major Accidents] (in Japanese). Tokyo: Koubundou, p. 141.

Medvejef, J. A. (trans. Yoshimoto, S.) (1992), Chernobyl no Isan [Chernobyl's Legacy] (in Japanese). Tokyo: Misuzu Shobo, p. 186, Note 323 (Webb, R.E., The health consequences of Chernobyl. *The Ecologist*, 16 (1986), pp. 169–170).

22 Report of the Aircraft Accident Investigation Commission, Ministry of Transport Japan; 19 June 1987, No. 62-2 (in Japanese).

23 *The Asahi Shimbun*, 8 October 1985 (evening edition).

24 *The Asahi Shimbun*, 8 October 1985 (evening edition).

25 *The Asahi Shimbun*, 4 September 1985.

26 *The Asahi Shimbun*, 3 September 1985.

27 *The Asahi Shimbun*, 6 September 1985.

28 *The Nihon Keizai Shimbun*, 2 November 1985.

All conversions from yen to US dollars in this text are based on exchange rates at the end of June 2003 ($1 = 122 yen).

29 *The Asahi Shimbun*, 6 September 1985.

In the Shizukuishi accident, on 30 July 1971, ANA flight 58 B-727-281 and the Self-Defence Forces training fighter North American F-86F crashed in midair above Shizukuishi, in Iwate Prefecture. In the Tokyo Bay Crash, on 9 February 1982, JAL flight 350 DC-8 crashed into Tokyo Bay through the actions of a mentally disturbed pilot.

30 *The Asahi Shimbun*, 4 September 1985 and 9 September 1985 (evening edition).

31 *The Asahi Shimbun*, 14 August 1985.

32 Ministry of Transport Japan, *'Koukuu Unyu Toukei Nenpyou' Koukuu*

Seisaku Kenkyuukai Kouseiken Series ['Air transport statistics annual report' Air Measures Research Institute Kouseiken Series] (in Japanese), No. 197, p. 59, material 12.

33 Report of the Aircraft Accident Investigation Commission, Ministry of Transport Japan (in Japanese), 19 June 1987, No. 62-2, p. 128.

34 Japan Flight Crew Union Federation (April 1994), Nikkou 123 bin ni kyuu gen'atsu wa nakatta [There was no rapid decompression on JAL flight 123] (in Japanese), p. 13.

   Kakuta, S. (1993), *Giwaku – JAL 123 bin tuiraku jiko* [Doubt – JAL123 Crash Accident] (in Japanese). Tokyo: Waseda Shuppan.

35 Report of the Aircraft Accident Investigation Commission, Ministry of Transport Japan, 62-2, Appendix, Koukuu Jiko Houkokusho Furoku JA8119 ni kansuru siken kenkyuu siryou [Aircraft accident report appendix: test research materials regarding JA8119] (in Japanese), p. 73, ref. additional diagram 4.

36 Aircraft Accident Report cited above, No. 62-2 (in Japanese), p. 101 3.2.2: 'Regarding the repair of damage resulting from the accident at Osaka Airport in 1978, and the operation and maintenance inspections of the aircraft that crashed'; p. 123 4.1.3: 'It was discovered that in the webbing between the upper and lower segments of the bulkhead, the edges and margins around the rivet holes were less than the values listed in the structural repair manual in some areas. This could have been caused by insufficient consideration for deformations in the back part of the fuselage during repair operations'.

37 Aircraft Accident Report cited above, p. 124, 4.1.3.7 (in Japanese): 'Based on this type of repair, the 1.18 connection segment that should have been connected by two rows of rivets were only connected by one row of rivets; the strength of this segment had dropped to about 70 per cent compared with the correct connection methods, and as a result this segment was likely to be in a condition that was susceptible to fatigue fractures'.

38 Report of the President's Commission on the Accident Three Mile Island (in English), *Three-mile tou genpatsu Jiko Houkokusho* [Three Mile Island Nuclear Accident Report] (1977) (in Japanese). Tokyo: Highlife Shuppan.

39 ICAO Accident Digest, Circular 47-AN/42 (16–45), Report of the public inquiry into the causes and circumstances of the accident that occurred on the 10 January 1954 to the Comet aircraft G-ALYP (in English).

   Job, M. (1997) *Air Disaster*, Vol. 1, p. 11 onwards. Australia: Aerospace Publications (in English).

   Gero, D. (trans. Shimizu, Y.) (1997), *Koukuu Jiko* [Aircraft Accidents] (in Japanese). Tokyo: Ikarosu Shuppan, p. 18 onwards.

40 ICAO Accident Digest, Circular 47-AN/42(16–45), Report of the public inquiry into the causes and circumstances of the accident that occurred on the 8 April 1954 to the Comet aircraft G-ALYY.

   Job, M., *Air Disaster*, Vol. 1. Australia: Aerospace Publications, p. 11 onwards.

Gero, D. (trans. Shimizu, Y.) (1997), *Koukuu Jiko* [Aircraft Accidents] (in Japanese). Tokyo: Ikarosu Shuppan, p. 18 onwards.

41 *Ibid.*, p. 19.

42 *Ibid.*

43 Republic of South Africa, Report of the Board of Inquiry into the Helderberg Air Disaster, ISBN0-621-13030-3, Appendix A.

44 Report of the Court Investigation India, Accident to Air India Boeing 747 aircraft VT-EFO, 'KANISHKA' on 23 June 1985, 26 February 1986.

45 NTSB AAR-90-01 (incomplete closure).

46 NTSB AAR-92-02, revised, adopted on 18 March 1992.

47 Report of the Aircraft Accident Investigation Commission, Ministry of Transport Japan, (1987.6.19) No. 62-2, Appendix (separate volume) (in Japanese), p. 31.

48 Belgium Commercial Airline Bureau, Accident Investigation Committee Report No. EW/A224 (in English), p. 25.

   JAL Flight Safety. August 1988 No. 59, Berugii Minkan Koukuukyoku Jiko Chousa Iinkai Jiko Houkokusho [Belgian Commercial Airline Bureau, Accident Investigation Committee Report No. EW/A224; abridged translation] (in Japanese), p. 47.

49 Aircraft Accident Report cited above, No. 62-2 Appendix (separate volume) (in Japanese), p. 31.

50 Belgian Commercial Airline Bureau, Accident Investigation Committee Report cited above, No. EW/A224 (in English), p. 29.

   JAL Flight Safety. August 1988, No. 59 (in Japanese), p. 47.

51 Report of the Court Investigation India, Accident to Air India Boeing 747 aircraft VT-EFO, 'KANISHKA' on 23 June 1985, 26 February 1986, p. 66 of Dr Hill's report.

52 Annex 13 of the Convention on International Civil Aviation concerning aircraft accidents and incident investigation first defines an 'accident' as an occurrence associated with the operation of an aircraft that takes place between the time any person boards the aircraft with the intention of flight until such time as all such persons have disembarked, in any of the following situations: (a) when a person is fatally or seriously injured as a result of being in the aircraft, or direct contact with any part of the aircraft, including parts that have become detached from the aircraft, or direct exposure to jet blast, except when the injuries are from natural causes (for statistical uniformity only, an injury resulting in death within 30 days of the date of the accident is classified as a fatal injury by the ICAO), self-inflicted, or inflicted by another person or hiding stowaways; (b) when the aircraft sustains damage or structural failure that adversely affects the structural strength, performance, or flight characteristics of the aircraft and would normally require major repair or replacement of the affected component (except for engine failure or damage, when the damage is limited to the engine, its cowlings or accessories, or for damage limited to propellers, wing tips, antennas, tyres, brakes, fairings, small dents or puncture holes in the aircraft skin); or (c) when the aircraft is missing or is completely inaccessible (an aircraft is

considered to be missing when the official search has been terminated and the wreckage has not been located).

An incident is 'an occurrence, other than an accident, associated with the operation of an aircraft that affects or could affect the safety of operation'. Then, among incidents, a serious incident is defined as 'an incident involving circumstances indicating that an accident nearly occurred'. Typical examples are appended as guidelines in the same Appendix.

Those examples include: aircraft structural failures or engine disintegration, multiple malfunctions of aircraft systems seriously affecting the aircraft, events requiring emergency use of oxygen, take-off or landing incidents such as undershooting, overrunning, or running off the side of runways, system failures, weather phenomena, and near collisions requiring an avoidance manoeuvre to avoid a collision.

Concerning these kinds of serious incident, it is recommended that an investigation be conducted by the state of occurrence (concerning abnormal situations and near misses, that the captain of the aircraft be obligated by the Civil Aeronautics Law of Japan to report to the Minister of Land, Infrastructure and Transport).

The International Nuclear Event Scale (INES) was established for the purpose of ensuring that a common understanding of unsafe events that have occurred in nuclear power facilities or in the transport of radioactive materials is shared by those involved in nuclear energy, mass media, and the general public. The scale has eight levels, ranging from 0 to 7, which evaluate an event on the basis of, most importantly, whether or not there has been release of radioactive material into the environment (contamination) and the scope of such contamination, whether or not people's bodies have been affected by it (exposure) and the degree to which they were affected, and whether or not there was safety defence in depth or damage to a reactor core and the degree of such damage. Events to which even the 0 level on this scale does not apply are considered 'events unrelated to safety' with respect to the INES evaluation and are not subject to evaluation.

Level 0 indicates an event that is 'not important with respect to safety', even though it may be an event that does affect safety, and is considered to be below scale.

Levels 1 to 3 indicate abnormal events. Level 1 events do not involve release of radioactive material into the environment and do not go beyond deviation from the approved operating range (deviation). Level 2 events involve considerable degradation of defence in depth, contamination through release of radioactive material on the facility premises only, or exposure of facility employees to radiation that exceeds the fixed annual limit (abnormal situation). Level 3 events involve a loss of defence in depth, the spreading of major contamination by on-premises release of radioactive material and acute injury to facility employees that results from exposure to radiation, release to outside the facility of a small amount of radioactive material that exceeds the permitted amount, or exposure of ordinary individuals to several tenths of a millisievert. Such events are regarded as serious incidents.

Events from levels 4 to 7 are accidents. Events of these levels all share the fact that radioactive material has been released to the outside, but are differentiated by the scope of the contamination and extent of exposure and by the degree of reactor core damage.

An event that affects safety-related facilities, such as equipment failure in a turbine system that results in a reactor shutdown, for example, but does not, in effect, result in release of radioactive material and does not affect the defence in depth, would be evaluated as a level 0 event according to the INES evaluation scale, below the scale by the INES criteria, and would not be regarded as an abnormal event. Such an event would, however, be regarded as an incident as we use the term. As a further example, even if there is a fire, which is generally considered to be an important accident, if it does not actually have any effect on the defence in depth, it is not subject to evaluation on the INES.

However, these events, too, are not only unsafe events that cannot be overlooked from the broad viewpoint of maintaining safety, they also have the potential to develop into INES level 1 or higher abnormal events or accidents through being linked to other events.

(In Japan, according to the Electric Utilities Industry Law and the Law for the Regulation of Nuclear Source Material, Nuclear Fuel Material and Reactors, incidents of INES level 0 and below that involve working nuclear power generation systems must be reported by the electric power company to the Minister of Economy, Trade and Industry. 'Minor problems', too, must be reported according to a notification from the Minister of Economy, Trade and Industry.)

From the above descriptions, we can see that the concept expressed by the term 'incident' as we use it in this book has a broader interpretation.

53 Aircraft Accident Digest (ICAO Circular 153-AN/56) pp. 22–68, 'KLM B-747, PH-BUF and Pan Am B-747 N736 collision at Tenerife Airport Spain on 27 March 1977', report dated October 1978, released by the Subsecretaria de Aviacion Civil, Spain, in both Spanish and English.

NTSB Identification DCA77RA014, 14 CFR General Aviation Form. Event occurred Sunday 27 March 1977 in Tenerife, Canr, Spain. Aircraft: Boeing 747, registration PH-BUP.

Job, M. (1997), Did he not clear the runway – the Pan American? *Air Disaster*, Vol. 1.1. Australia: Aerospace Publications (in English), p. 164 onwards.

54 Report dated October 1978 released by the Subsecretaria de Aviation Civil, Spain. Part one Comments of the Netherlands Department of Civil Aviation.

55 Job, M. (1997), Ref. *Air Disaster*, Vol. 1, p. 174. Australia: Aerospace Publications (in English).

56 NTSB AAR76-12, adopted on 31 March 1976.

57 NTSB AAR73-15, adopted on 5 July 1973.

ANA 1981, *Koukuu Jiko Digest* [Airline Accident Digest] (in Japanese), Part 3, p. 19.

58 See note 54 cited above.

59 NTSB AAR-75-09, adopted on 23 May 1975.

60 Report of the Aircraft Accident Investigation Commission, Ministry of Transport Japan, No. 60-6 (in Japanese).
ANA, *The Montage of Aircraft Accidents* (in Japanese), Part 1, p. 3 onwards.

61 Aircraft Operating Manual, DC-10 78, Exhaust No. 3-1, Thrust reverser (in Japanese).

62 Aircraft Operation Manual, AOM, Ch. 3, Table 24 (in Japanese).

63 DC-10 Maintenance Manual 78-00-00 (in English), Exhaust General, Maintenance Practices 202.

64 Report of the Aircraft Accident Investigation Commission, Ministry of Transport Japan, 1 May 1983, No. 58-3.

65 Japanese Criminal Law, Article 211.

66 The law related to punishment for actions, etc., that result in danger to aircraft; Article 6, Para. 2.

67 The Civil Aeronautics Law of Japan, Article 30 (Revocation, etc., of competence certificate).

68 Lederror, J., 'Airmen have always considered an accident or an incident as an opportunity to learn. Some call this tombstone safety. The antithesis of tombstone safety is system safety which may be defined loosely as putting your hand right where your foresight should be'.

69 See note 38 cited above.

70 NTSB AAR 79-17, adopted on 21 December 1979.
ANA Flight Safety Review (in Japanese), No. 90, p. 63 onwards.

71 Report of the Aircraft Accident Investigation Commission, Ministry of Transport Japan (in Japanese), 25 September 1985, No. 60-5.

72 Report of the Aircraft Accident Investigation Commission, Ministry of Transport Japan (in Japanese), 28 October 1988, No. 63-9B.

73 *The Asahi Shimbun*, 7 December 1985.

74 Report of the Aircraft Accident Investigation Commission, Ministry of Transport Japan (in Japanese), 19 July 1996, No. 96-5.

75 No public organizations were involved in the internal investigations related to the incident involving the Airbus 300-600 that occurred on 1 March 1985, and not all of the information obtained has been made publicly available.
The incident involving Finland Airlines A300 B4-203FF that occurred on 9 January 1989, during approach to Helsinki Airport, was the subject of an investigation by Finland's public investigation organization, and was discussed at the International Society of Air Safety Investigators (ISASI) seminar the following year, in 1990. The incident involving an A310 during approach to Moscow Airport on 11 February 1991 was also the subject of an investigation by a public investigation organization, and was written up in a January 1992 publication of the Flight Safety Foundation (FSF) (cited in above Report 74, pp. 261–2).

76 National Research Council (1997), *Aviation Safety and Pilot Control.* Washington, DC: National Academic Press (in English), p. 24.

77 *The Nihon Keizai Shimbun,* 30 July 2004, evening edition.

'Regarding this accident, the Captain had been suspected of professional negligence resulting in death or injury. But on 30 July 2004, the Nagoya District Court passed down a verdict of 'not guilty', denying negligence on the grounds that the Captain's operation of the aircraft contributed to the accident, but he could not foresee that his actions would lead to an accident resulting in death or injury.'

78 Report of the Aircraft Accident Investigation Commission, Ministry of Transport Japan (in Japanese), 17 December 1999, No. 99-8, pp. 22–23.

79 *Ibid.*, p. 37, 3.3.4.

80 NTSB Identification MIA94FA169, 14 CFR 121 operation of American Airlines, Inc. Accident occurred Wednesday 29 June 1994 in Caribbean, CB. Aircraft: McDonnell Douglas MD-11, registration N1752K.

81 The objective validity of MAIR, which was newly developed as a method for analysing incidents related to human factors, has been verified using quantification method III by extracting only the information source, as well as the error modes in the information receiving, decision, and action/ instruction phases, from the results of the analysis of incidents reported by the crew, as will be discussed later. As shown in the figure, the error modes for each category are distributed close together, proving that they have similar characteristics, and proving the objective validity of this analysis method. What is deserving of attention, however, is that factors in the seventh category – Preconception in the information receiving phase, unreasonableness/high-handedness in the decision phase, and recklessness/ forcefulness in the action/instruction phase – are in close proximity on the X axis but are far apart on the Y axis. This is because of the effects of correlations with factors related to the information source; namely weather information, outside visual information, ATC advice, instruments, instructions/operations by the captain, and charts/manuals. This also supports the claim that improvements in hard elements are an urgent issue in terms of reducing the degree of danger.

High values have been derived for the square roots of eigenvalues, which are said to express the accuracy of the analysis:

First dimension: 0.8197
Second dimension: 0.7648
Third dimension: 0.7203

82 Okinaka Shigeo Sensei wo Shinobu Kai Henshu [Published in memory of Dr Shigeo Okinaka] (1992). *Okinaka Shigeo – I no michi* [Shigeo Okinaka – the Way of Medicine] (in Japanese). Tokyo: Japan Medical Journal, pp. 60–62.

83 Mori, M. (Ed.) (1984), *Nigai Karute – Shourei ni Manabu Goshin Yobougaku* [Bitter Charts – Misdiagnosis Prevention as Learned from Actual Cases] (in Japanese). Tokyo: Nikkei Mcgraw-Hill.

Difficult operation
(action/instruction)

Delay (Information

Could not see/hear, could not
see clearly/hear clearly• Unreasonableness/
• high-handed
• Vacillation (decision)

Loss of stability• • Vacillation (Information
receiving)

• Weather information

Recklessness, forcefulness,
excessive operation •

Easy-going, assumption • • Outside visua

ATC instructions   •
Failure to see

Crew communications
(captain's commands and   •
directives, other crew voice)

Documents
• (chart/manual, etc.)

Wrong conviction/   Incorrect action/instructic
hasty conclusion/   •inappropriate action/ inst
one's own interpretation

Saw incorrectly/ • Preconception
heard incorrectly/
mistaken impression

Verification of appropriateness of MAIR (see colour plate section)

receiving)

 • Delay (action/instruction)
Delay (decision)/insufficient consideration

 • Fixation, persistence action

 • Fixation

I information   • Excessive focus

 • Instruments
_____

                                    Forgot •   •
                                    (no action/instruction)   Forgot
ɘ/hear/notice   (no decision)

ɔn,
ruction

$\sqrt{\text{eigen value}}$
     1st dimension: 0.8197
     2nd dimension: 0.7648
     3rd dimension: 0.7203

Chapter 2

1 Miyao, K. (1994), Hirou to Shikinou [Fatigue and visual function]. *Igaku to Kougaku kara mita Koutsuu Anzen Taisaku* [Traffic Safety Measures as Seen from Medicine and Engineering] (in Japanese), Vol. 1994, pp. 13–20.

2 Fukuda, T. (1991), Koureisha no Shikaku Kinou [Ageing and visual functions] (in Japanese). *IEICE Technical Research Report*, Vol. 90, No. 409, pp. 1–8.

   Nishida, S. (1989), Gansoshiki no rouka to chousetsu [Ageing changes of ocular tissues and their influences on accommodative functions] (in Japanese). *Japanese Ophthalmological Society Journal*, Vol. 94, No. 2, pp. 93–119.

   Tokoro, T. (1986), Gan Kinou no Nenrei Henka [Changes in visual functions due to age]. *Shiryoku zukai rinshou ganka kouza: roujin to me* [Vision: an Illustrated Course on Clinical Opthamology: The Elderly and Eyes] (in Japanese). Tokyo: Medical Review Co., Ltd, pp. 16–17.

3 Report of ICAO Fact-Finding Investigation C-WP/7764.1983.12, Destruction of Korean Air Lines Boeing 747 over sea of Japan, August 31 1983.

   Minutes of the 101st National Assembly House of Councillors Committee on Budget (23 February 1984).

4 There are assumptions that, because of the proximity to the North Pole, although it was necessary to change from the magnetic compass to the gyroscope, the aircraft was still flying by magnetic compass, and that, furthermore, although it should have been easy to notice this error, the aircraft was being flown without notice of this error. Given this situation, the only possible assumption is that the pilot had fallen asleep while flying.

5 Moore-Ede. M. (trans. Aoki, K.) (1994), *Daijiko wa yoake-mae ni okiru* [Major Accidents Happen before Dawn] (in Japanese). Tokyo: Kodansha, p. 158.

6 Annex 6 of the Convention on International Civil Aviation, Part 1, 4–5, duties of pilot-in-command: 'The pilot-in-command shall be responsible for the safety of all crew members, passengers, and cargo on board when the doors are closed, and shall also be responsible for the operation and safety of the aeroplane from the moment the aeroplane is ready to move for the purpose of taking off until the moment it finally comes to rest at the end of the flight and the engine(s) used as primary propulsion units are shut down'.

   Annex 2 of the Convention on International Civil Aviation, 2–4, stipulates the following regarding the 'Authority of pilot-in-command of an aircraft': 'The pilot-in-command of an aircraft shall have final authority as to the disposition of the aircraft while in command'.

7 Convention on Offences and Certain Other Acts Committed on Board Aircraft, Article 6 (restraint), Article 8, Para. 1 (deplaning), Article 9, Para. 1 (handing over).

   Civil Aeronautics Law of Japan Article 73-3 (measures, etc., to suppress acts that impede safety of pilot in command).

8 Report de la commission d'enquête sur l'accident survenu le 20 janvier 1992

près de Mont Sainte-Odile (Bas Rhin) à l'Airbus A320 immatriculé F-GGED exploité par la compagnie Air Inter F-ED 920/120 (in French).

ANA, *The Montage of Aircraft Accidents* (Okano, M., Ed.) (1997), Vol. VII, Glass cockpit (in Japanese), pp. 104–105.

Gero, D. (trans. Shimizu, Y.) (1997), *Koukuu Jiko* [Aircraft Accidents] (in Japanese). Tokyo: Ikarosu Shuppan, p. 226.

9 When CP Air Flight 402 (DC-8-43) was landing at Tokyo International Airport, it descended at a sharp downward angle in an attempt easily to see the runway which was subject to diffused reflection in fog; the bottom of the aircraft hit the sea protection wall, cleared the wall, and ran uncontrolled to the end of the runway, where it broke apart and burst into flames. Accident Investigation Report, Canadian Pacific Airline DC-8, CF-CPK Tokyo International Airport, 4 March 1966; Canadian Pacific Airline and BOAC Aircraft Accident Technical Investigation Commission, 4 March 1969 (in Japanese).

This accident is said to be an example of a 'duck under' accident.

10 Ueda, T. (1988), Raiun no Katsudo [Activities of thunderclouds] (in Japanese). *Aviation Engineering*, No. 396. Tokyo: Japan Aeronautical Engineer's Association, p. 35.

11 NTSB AAR-72-12, adopted on 29 December 1971.

ANA Koukuu Jiko Digest [ANA Air Accident Digest] (1981), Part 3: Approach and landing (in Japanese), p. 137 onwards.

12 NTSB AAR-76-12, adopted on 31 March 1976.

13 NTSB AAR-79-12, report dated 16 August 1979, *ANA Koukuu Jiko Digest* [ANA Air Accident Digest] (1980), Part 1: Take off (in Japanese), p. 170 onwards.

14 NTSB AAR-82-08, adopted on 10 August 1982.

15 Through Japanese administrative reforms in January 2001, the Ministry of Transport has merged with the Ministry of Construction to become the Ministry of Land, Infrastructure and Transport.

16 NTSB AAR-76-08, adopted on 12 March 1976, *ANA Koukuu Jiko Digest* [ANA Air Accident Digest] (1981), Part 3: Approach and landing, p. 245.

17 National Academy of Sciences (1983), *Low Altitude Wind Shear and its Hazard to Aviation*. Washington, DC: The National Academic Press, pp. 14–15.

18 Report of the Aircraft Accident Investigation Commission, Ministry of Transport Japan (in Japanese), 25 September 1985, No. 60-5.

19 According to the Airport Development Law, public airports provided for use in air transport are divided into one of three classes, which are stipulated by government decree. Class 1 refers to airports required for international air transport services, and includes Tokyo International Airport, New Tokyo International Airport, Osaka International Airport, and Kansai International Airport. Class 2 refers to airports required for primary domestic air transport services, and Class 3 refers to airports required for securing regional air transport services.

20 Gero, D. (trans. Shimizu, Y.) (1997), *Koukuu Jiko* [Aircraft Accidents] (in Japanese). Tokyo: Ikarosu Shuppan, p. 187.
21 NTSB AAR-91-05, adopted on 25 June 1991.
22 Many airports in the United States are equipped with wind shear detection devices. Improvements have since been made in Japan as well, with Doppler radars being installed since 1995 at Kansai International Airport, New Tokyo International Airport, the new Chitose Airport, and Osaka International Airport, and with plans for installation at Naha Airport in 2003.
23 NTSB AAR-76-15, adopted on 19 May 1976.
24 NTSB AAR-91-04, adopted on 30 April 1991.
25 NTSB AAR-73-14, adopted on 14 June 1973.
    ANA Koukuu Jiko Digest [ANA Air Accident Digest] (1981), Part 3: Approach and landing. Tokyo: All-Nippon Airways Co., Ltd, p. 11.
26 Japan Aircraft Pilot Association (1985), *Pilot* (in Japanese), No. 5, p. 33 onwards.
27 *ANA Koukuu Jiko Digest* [ANA Air Accident Digest] (1980) (in Japanese), Part 1: Take off. Tokyo: All-Nippon Airways Co., Ltd, p. 87.
    ANA *The Montage of Aircraft Accidents* (1990) (in Japanese). Tokyo: All-Nippon Airways Co., Ltd, p. 158 onward.
28 Ministry of Transport, Civil Aviation Bureau, 5 August 1970, Aircraft Investigation No. 68 (in Japanese).
    NTSB Identification SEA69A0062, 14 CFR General Aviation Form. Event occurred Tuesday, 24 June 1969 in Moses Lake, WA. Aircraft: Convair 880, registration JA-8028.

Chapter 3

1 In Japan, following this research, from January 2001, aircraft provided for air transport operations flying in a control zone or a control area are required to have TCAS installed, according to the Civil Aeronautics Regulation of Japan, Article 147, Subpara. 5. However, TCAS is not required on small aircraft with less than 30 passenger seats, or with a maximum take-off weight of less than 15 000 kg. Part 1, 6–18 of Annex 6 to the Chicago Convention states that, as of 1 January 2003, all aircraft with 30 or more passenger seats and weighing 15 000 kg or more at take-off shall be equipped with ACAS II.
2 Special feature, *The Tokyo Shimbun*, 24 December 1988 (Top ten news items of 1988, as selected by readers), 'The former captain of the submarine *Nadashio* and the former captain of the fishing vessel *Dai-ichi Fujimaru* were charged with professional negligence causing death and professional negligence interfering with traffic'.
    10 December 1992: 'Negligence was recognized on the part of both captains; a prison term was handed down with a suspended sentence, and the decision has been finalized'.

Chapter 4

1 NTSB AAR-89-03, adopted on 14 June 1989.
  JAL Flight Safety, February 1990, No. 68, Testimony by Dr Colen Drury.
2 Oshima, M., Yamamoto, S., and Yokobori, S. (1967), *Koukuu Igaku* [Airline Medicine] (in Japanese). Tokyo: Igaku Shoin, p. 157.
3 Japan Aeronautical Engineer's Association Aviation Engineering (1989) (in Japanese), No. 412, p. 12.
  Newhouse, J., Ishikawajima-Harima Heavy Industries Co., Ltd PR Div. (Ed.); Aircraft Industry Research Group (trans.), *Sporty Game* (1988). Tokyo: Gakuseisha.
4 AIB (UK Accident Investigation Board) Aircraft Accident Report 78 AAR-78-9.

Epilogue

1 Hirono, R. (1996), *Sabakareru Yakugai Aids* [Judgement on Drug-induced AIDS] (in Japanese). Tokyo: Iwanami Shoten, Iwanami Booklet No. 417, p. 22.
2 *The Asahi Shimbun*, 25 August 1995.
  The Nihon Keizai Shimbun, 10 March 1997.
3 *The Asahi Shimbun* (evening edition), 23 June 1995.
4 *The Asahi Shimbun*, 30 March 1996.
5 Hirono, R. (1996), *Sabakareru Yakugai Aids* [Judgement on Drug-induced AIDS] (in Japanese), p. 8 (Highlights of Tokyo District Court), Iwanami Booklet No. 417. Tokyo: Iwanami Shoten.
6 Shibata, T. (1994), 'Kagaku Houdou' 4-2, Thalidomide-ji Hassei Kyuuzou [Science News: Rapid increase in thalidomide babies] (in Japanese), *The Asahi Shimbun*, pp. 11–13.
7 Editorial, *The Asahi Shimbun*, 23 February 1989.
8 Tensei Jingo (Vox Populi, Vox Dei), *The Asahi Shimbun*, 25 January 1996.
9 *The Asahi Shimbun* (evening edition), 11 March 1988: Choukikan Tairyou Touyo [Long-term, high dosage administration], *The Asahi Shimbun*, 11 and 12 March 1988.
10 *The Asahi Shimbun*, 11 March 1988.
11 *The Asahi Shimbun*, 11 March 1988.
12 Tensei Jingo (Vox Populi, Vox Dei), *The Asahi Shimbun*, 12 March 1988.
13 *The Asahi Shimbun* (evening edition), 11 March 1988.
14 *The Asahi Shimbun* (evening edition), 11 March 1988.
15 *The Asahi Shimbun*, 12 March 1988.
16 *The Asahi Shimbun*, 12 March 1988.
17 *The Asahi Shimbun*, 12 March 1988.
18 Harada, M. (1972), *Minamata-byou* [Minamata Disease] (in Japanese). Tokyo: Iwanami Shoten, p. 2 onwards.
19 *Ibid.*, p. 6.
  Ibid., p. 48; it was reported in *Lancet* that this was extremely similar to the

organic mercury poisoning of workers at an agricultural chemical plant that occurred in England in 1940, as reported by Hunter Russell (September 1958).

20 Harada, M. (1972), *Minamata-byou kara Manabu* [Learning from Minamata disease].

    Yasuda Kasai Kaijou – Kankyou Kouza [Yasuda Fire and Marine Insurance Company, Limited – Environment Course] (in Japanese). Tokyo: Chouhoki, p. 84.

21 *The Asahi Shimbun* (1995), 'Chiezo', p. 339.

22 Firefighting activities in the case of the fire at the Power Reactor and Nuclear Fuel Development Corporation Tokai plant in March 1997 lasted for only 1 min, but water spray lasting over 8 min is said to have been required in the fire at a similar facility in Belgium in 1981.

    Section 5: Accidents, *The Ibaraki Shinbun*, 27 May 2000.

23 Iguchi, T. (1999), *Kokuhaku* [Confession] (in Japanese). Tokyo: Bungeishunju. The cumulative total losses over 11 years (from 1984 to 1995) in the case of the cover-up of losses in US Treasury trading in 1995 by a Daiwa Bank trader are said to be over 1.1 billion dollars.

24 Barings Bank: on 26 February 1995, a trader covered up losses in futures trading. Losses totalled 927 million pounds (1.5 billion dollars).

    Hiwatari, J. and Ashida, H. (2002), Operational Risk Kanrino Koudo-ka ni kansuru ronten seiri to kongo no kadai – teigi-teki risk kanri houhou dounyuu e no torikumi wo chuushin ni [Summary of key points and future issues related to the improvement of operational risk management – focusing on efforts targeting the introduction of defined risk management methods] (in Japanese), Bank of Japan Assessment Bureau (2002), Discussion Paper No. 02-J-1, p. 4.

    Leeson, N. (trans. Toda, H.) (1997), *Watashi ga Barings Bank wo Tsubushita*. Tokyo: Shinchousha [Nick Leeson 'Rogue Trader'].

25 NTSB AAR-75-16, adopted on 26 November 1975.

26 Minutes of the 112th National Assembly House of Councillors Committee on Audit (25 April 1988).

Index